Motor Vehicle Fuel Economy

Other books by the same author

Introduction to Internal Combustion Engines
Management of Engineering Projects (editor)

Other Macmillan titles of related interest

Mechanical Reliability, second edition
A. D. S. Carter

Elementary Engineering Mechanics
G. E. Drabble

Principles of Engineering Thermodynamics, second edition
E. M. Goodger

Aviation Fuels Technology
E. M. Goodger and R. A. Vere

Polymer Materials: An Introduction for Technologists and Scientists
Christopher Hall

Gas Turbine Engineering: Applications, cycles and characteristics
Richard T. C. Harman

Introduction to Engineering Materials, second edition
V. John

Strength of Materials, third edition
G. H. Ryder

Mechanics of Machines
G. H. Ryder and M. D. Bennett

Engineering Heat Transfer, second edition
J. R. Simonson

Turbocharging the Internal Combustion Engine
N. Watson and M. S. Janota

Essential Solid Mechanics–Theory, worked examples and problems
B. W. Young

Motor Vehicle Fuel Economy

Richard Stone
Brunel University
Uxbridge, Middlesex

MACMILLAN

First published 1989

Published by
MACMILLAN EDUCATION LTD
Houndmills, Basingstoke, Hampshire RG21 2XS
and London
Companies and representatives
throughout the world

Printed in Hong Kong

British Library Cataloguing in Publication Data
Stone, Richard, *1955–*
Motor vehicle fuel economy.
1. Motor vehicles—Fuel consumption
I. Title
629.2′53 TL151.6

ISBN 0–333–43820–5

Contents

Preface

Motor vehicle fuel economy is of personal interest to all engineers who drive cars, and of professional interest to the many engineers employed in the automotive industry. This book is aimed at final year students who are perhaps specialising in automotive engineering, or young graduates who are employed in the automotive industry.

Significant improvements in fuel economy have been made recently, and this is reflected in the large number of papers published in journals and conference proceedings. However, even collected papers do not provide a coherent overview of motor vehicle fuel economy, and they may pre-suppose a knowledge that is absent. Consequently, this book aims to provide a self-contained treatment of the factors that affect fuel economy, and how fuel economy is optimised.

The strategy for reducing fuel consumption has two main aspects: to provide the power required for propulsion more efficiently, and to reduce the power required for propulsion. These arguments are developed in chapter 1, and the factors that influence the fuel consumption in the real world are discussed. Spark ignition and diesel engines are discussed in chapters 2 and 3, with particular attention to the factors that govern their efficiency. Different transmission systems are discussed in chapter 4, along with powertrain matching, and optimisation for performance and fuel economy.

The factors that reduce the propulsive power requirement are discussed in chapters 5 and 6. Chapter 5 is devoted to vehicle aerodynamics, covering the topics of drag reduction, lift control and stability. The effect of drag reduction on performance and fuel economy is treated quantitatively. The rolling resistance is mostly dependent on the tyres and the vehicle mass, both of which are treated in chapter 6. Techniques for reducing the vehicle mass are particularly important in the real world, since vehicles do not travel at a constant speed on horizontal ground. Two case studies are considered in the final chapter–the Rover 800 car and the Ford Transit van. Both vehicles are analysed, to illustrate how the techniques described in the first six chapters lead to a low fuel consumption.

Throughout the book, attention is drawn to the use of micro-electronics and computing, to illustrate their contribution towards producing more fuel efficient vehicles. At the end of each chapter there are discussion points; these should enable any reader to check his or her understanding of the preceding material.

Any book owes a lot to many people. Those who deserve individual thanks are Mr J-P Pirault, Dr M L Wyszynski and Dr Neil Richardson for reading and commenting on the draft, and Ruth Sterland for the typing and inevitable retyping.

Summer 1987 RICHARD STONE

Acknowledgements

The author and publisher wish to thank the following, who have given permission to use copyright material:

The Society of Automotive Engineers, Inc. for figures 1.2, 2.14, 3.5, 3.13, 4.18, 5.14, 5.15, 5.16 and 5.17.
Department of Energy for figure 1.7.
Bedford Vehicles for figure 1.9.
Chloride Silent Power Ltd for figure 1.10.
T & N plc for figure 2.7.
Council of the Institution of Mechanical Engineers for figures 2.10, 2.12, 3.11, 3.12, 3.14, 3.15, 4.13, 4.14, 4.15, 4.19, 4.20, 5.6, 5.21, 6.8, 7.9 and 7.10.
Johnson Matthey for figure 2.13.
Ford Motor Company for figures 3.3, 4.6, 4.9, 4.12, 6.3, 6.4, 7.7 and 7.8.
Thorsons Publishing Group for figure 3.7.
Butterworth and Co. (Publishers) Ltd for figure 4.11.
Design Engineering for figure 4.16.
Plenum Publishing Corporation for figures 5.3, 5.4, 5.5, 5.19 and 5.24.
Austin Rover Group Ltd for figures 7.1 and 7.2.

Material is acknowledged individually throughout the text of the book. Every effort has been made to trace all the copyright holders but if any have been inadvertently overlooked the publisher will be pleased to make the necessary arrangement at the first opportunity.

Notation and Terminology

Notation

A	vehicle frontal area (m^2)
A_r	radiator area (m^2)
abdc	after bottom dead centre
atdc	after top dead centre
b	breadth (m)
bbdc	before bottom dead centre
bmep	brake mean effective pressure (N/m^2)
btdc	before top dead centre
C_d	drag coefficient (equation 5.1)
C_{dr}	radiator flow drag coefficient
C_l	lift coefficient (equation 5.1)
C_R	tyre rolling resistance coefficient
CAD	computer-aided design
CAE	computer-aided engineering
CAFE	corporate average fuel economy
CAM	computer-aided manufacture
CO	carbon monoxide
CTX	continuously variable transaxle
CVT	continuously variable transmission
d	characteristic dimension (m)
D	drag force (N), cylinder diameter (m)
D_d	drum diameter (m)
D_f	friction drag (N)
D_p	pressure drag (N)
D_r	radiator drag (N)
D_t	tyre diameter (m)
DI	direct injection
ECE	Economic Commission for Europe
EEC	European Economic Community

EGR	exhaust gas recirculation
EPA	Environmental Protection Agency (US)
F	force (tractive) (N)
FEM	finite element method
FMS	flexible manufacturing system
FTP	federal test procedure
g	acceleration due to gravity
gr	gearing ratio (engine speed to vehicle speed)
grp	glass fibre reinforced plastic
h	height (m)
HC	unburnt hydrocarbons
HRCC	high compression ratio compact combustion chamber
I	moment of inertia (kg m^2)
IDI	indirect injection
l	length (m)
L	stroke (m)
M	Mach number, mass (kg)
M_{eff}	effective mass (kg) (equation 5.5)
m_f	mass flow rate of fuel (kg/s)
MON	motor octane number
mpg	miles per gallon
mph	miles per hour
N'	engine speed, revolutions per second, divided by 2 for 4-stroke engines (equation 2.2)
NO_x	nitrogen oxides
p	tyre inflation pressure (N/m^2)
p_b	brake mean effective pressure (N/m^2) (equation 2.2)
ppm	parts per million (number of carbon atoms)
Q_r	air flow through the radiator (m^3/s)
r	radius of curvature, road wheel radius (m)
R	rolling resistance (N)
R_{td}	tyre rolling resistance measured on a drum (N)
R_{tf}	tyre rolling resistance measured on a flat surface (N)
r_v	volumetric compression ratio
RON	research octane number
sfc	specific fuel consumption (for example, kg/kWh)
SI	spark ignition
T	torque (N m)

TSI	timed sequential injection
v	vehicle velocity (m/s)
v_a	air velocity (m/s)
v_r	air velocity through the radiator
V_s	swept volume or displacement (m^3)
$\dot{W}$	power (kW)
$\dot{W}_b$	engine output or brake power (kW)
$\dot{W}_f$	power dissipated by friction (kW)
$\dot{W}_w$	power at the driving wheels (kW)
α	yaw angle, diesel engine load ratio
β	angle of incidence
γ	ratio of gas specific heat capacities
η	efficiency
θ	inclination of the road from the horizontal
μ	dynamic viscosity (Ns/m^2 or kg m/s)
ρ	density (kg/m^3)
τ_w	wall shear stress (N/m^2)
ϕ	equivalence ratio, local inclination of a surface
ω	angular velocity (rad/s)

Terminology

Fuel consumption is used here to mean the reciprocal of *fuel economy. Fuel economy* (in miles per gallon or kilometres per litre) is analogous to the efficiency of an engine. *Fuel consumption* (for example, as litres per 100 km) is the equivalent of the specific fuel consumption of an engine (kg/kWh or kg/MJ). The *road load* is the torque required of an engine that corresponds to a particular *tractive force, resistance* or *effort.*

1 Introduction

1.1 The strategy for improved fuel economy

The strategy for reducing vehicle fuel consumption is very straightforward, with just two aspects: firstly to reduce the power required to propel a vehicle, and secondly to produce the power required for propulsion more efficiently. The power required to propel a vehicle is governed by the aerodynamic resistance, which is the subject of chapter 5, and by the rolling resistance. A vehicle's rolling resistance is dependent on the vehicle weight and the rolling resistance coefficient of the tyres; both of these issues are addressed in chapter 6 on vehicle design.

Whether the motive power comes from a spark ignition engine (the subject of chapter 2) or a diesel engine (chapter 3), the fundamental requirement is for efficient operation, especially at part load. The power has to be delivered to the driving wheels as efficiently as possible, and careful matching of the transmission ratios is also needed; this is to ensure that the engine is working at an efficient operating point. Powertrain optimisation is discussed in chapter 4, where different transmission systems are also described.

Improved fuel economy has to be achieved within cost constraints, and due account also has to be taken of performance, styling, emissions requirements, reliability and maintainability. The subject of exhaust emissions is particularly important; it is discussed further in this chapter, and in chapters 2 and 3. Fortunately, there is no unique optimum solution for vehicle design, and there will continue to be a wide range of vehicles in each category. This provides engineers with a welcome challenge, in meeting the needs of different users and markets.

In the real world, vehicles do not travel at constant speeds on horizontal ground with no wind. Vehicle fuel economy will also be affected by the traffic flow, the driver, the vehicle condition and whether or not the engine is fully warmed-up. Attitudes to fuel economy are also affected by fuel costs, and these are the subject of wide international variation, which is largely due to different taxation policy. Fuel costs are also directly affected by the cost of crude oil, and this too is subject to variations caused by government's actions.

Table 1.1 Performance data for the Ford Escort [from Shell (1986)]

Model year	Power (kW)	Top speed		Fuel consumption at 120 km/h	
		(km/h)	(mph)	(litres/ 100 km)	(mpg)
1969	45	137	86	9.1	31.2
1985	51	157	98	7.1	40.0

The value that can be assigned to fuel economy savings also varies; it is shown in section 1.4 that on a fixed mileage basis, initial improvements have a greater significance than subsequent improvements. Furthermore, the law of diminishing returns applies, since the simplest, cheapest and most effective measures for improving fuel economy are usually applied first. None the less, improvements in vehicle performance are continuing to be made, and this is illustrated by the data in table 1.1 for the Ford Escort.

These trends are typical of those exhibited by other vehicles and manufacturers.

Improved fuel economy is unlikely to come from the use of new forms of thermal power plant; reciprocating internal combustion engines are likely to continue in use for the forseeable future. In the past, electric vehicles with lead acid batteries have been successful for a limited range of applications. The improved energy storage offered by sodium sulphur batteries will enable electric vehicles with improved performance to be developed. However, it will be shown in section 1.5 that when due account is taken of the efficiency of electricity production, distribution, storage and conversion, the overall efficiency and operating costs are comparable to those of liquid-fuelled vehicles.

The remainder of this chapter examines how fuel economy is effected in the real world, and whether or not other forms of motive power are feasible.

1.2 Driving patterns and the owner/driver influence

Vehicle driving patterns are governed by the attitude of the driver and external constraints. The driver is the biggest variable over which the engineer has no control. Inevitably, a vehicle driven with gentle acceleration and minimal braking will use less fuel than a vehicle which is continually being accelerated and braked. The driver also has a responsibility towards the correct maintenance of the vehicle. A survey reported by Atkinson and Postle (1977) shows a surprisingly high incidence of maladjustment on spark ignition engines submitted for tuning; the results are shown in table 1.2. The averaged reduction in fuel economy from controlled malfunctions on a range of engines is shown in table 1.3.

The current use of electronic ignition and fuel injection systems will eliminate many of the problems reported in table 1.3. However, fuel injectors are also subject to ageing. The further a vehicle travels, the more 'run-in' the reciprocating and rotating components become, and the lower the frictional losses. In a properly maintained vehicle, this leads to an increase in fuel economy with elapsed mileage, as shown in figure 1.1. These results

Table 1.2 Incidence of parameters outside manufacturer's specification on a sample of 72 vehicles

Correction needed	Percentage of sample
Mixture strength at idle	83.4
Static ignition timing	75.0
Dwell angle (contact breaker gap)	40.6
Valve gear adjustment	29.2
Spark plug replacement	23.6
Contact breaker replacement	20.8
Mixture strength at 2000 rpm	18.1
Cylinder leakage	16.7
Air cleaner replacement	5.6

Table 1.3 Effect of artificially introduced engine maladjustments on fuel economy

Factor	Percentage reduction in fuel economy
One failed spark plug	12.6
Mixture strength from weak to rich	10.5
Idling speed increased from 650 to 850 rpm	5.1
Seized centrifugal advance mechanism	13.4
Failed vacuum advance advice	2.8
Removal of thermostat during warm-up	2.1
Restricted air cleaner element	11.5

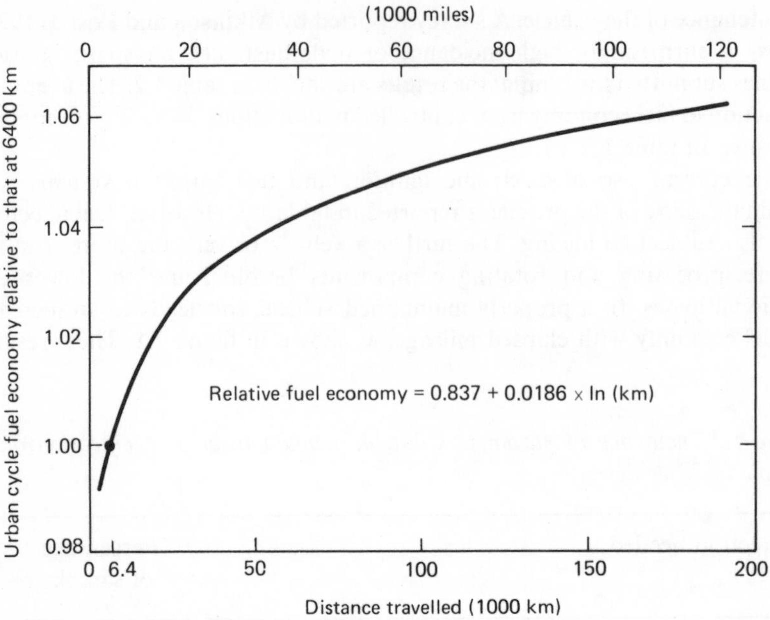

Figure 1.1 The improvement in vehicle fuel economy with the distance travelled, adapted from Murrell (1979)

from Murrell (1979) are applicable to all cars, since no significant dependence was found for either the manufacturer, the model year or the vehicle weight.

Finally, vehicle maintenance extends beyond the engine, for example, the data presented in section 6.2 imply that if the tyre inflation pressure is 10 per cent low, then there will be a 1 per cent penalty in fuel economy for cars, and a greater penalty for trucks and buses.

External factors affecting the driver are the traffic flow patterns and speed limits (these are often used in an attempt to save fuel, as well as to improve safety). Berg (1985) reports that, in German city traffic, idling can account for 35–45 per cent of the time. This is obviously undesirable for both fuel economy and emissions. Traffic flow can be improved by better roads, the improved design of junctions, and traffic management systems in which the traffic lights at adjacent junctions are sequenced.A study in a West German city reported by Lohr (1984) attributed a 20 per cent fuel saving to improved traffic flow.

Fuel economy devices are another area where the savings can be attributed to the driving pattern. Even if a fuel economy device shows no improvement in fuel economy when an engine is tested with a dynamometer, there can still be a reduction in vehicle fuel consumption. When such a device is fitted the engine will probably be tuned and, furthermore, the driver may drive more carefully (perhaps subconsciously), with the net result that fuel is saved. In other words, fuel economy devices can lead to fuel savings because of indirect effects.

1.3 Real world fuel economy and driving cycles

The problem with measurements of vehicle fuel economy is that the most reproducible results (those from driving at constant speed under controlled conditions) are those that are least typical of normal usage. Conversely, the conditions that are typical of normal use (say city driving) are the least reproducible. Some techniques for establishing comparative fuel consumption data in the real world are discussed by Burt (1977). In order to have the reproducible results that are essential for comparisons between vehicles, both specified driving cycles and constant speed tests are used. The driving cycles are also used for assessing exhaust emissions, and the different international test procedures are detailed by Barnes and Donohue (1985). The emissions requirements for the European Economic Community (EEC) are given in table 2.2. The European urban driving cycle ECE 15 is shown in figure 1.2. Results presented by Jarvis (1984) for this urban cycle show that 26 per cent or more of the total fuel consumed can be attributed to periods of idling and over-run. Over-run occurs during deceleration, when the throttle has been closed, and the engine is being 'driven' by the vehicle. This leads to various approaches for improving the urban cycle fuel economy, including:

(a) Electronic control of idling to minimise the engine speed. Ma (1986) suggests that a 200 rpm reduction in the engine idling speed could lead to a 6.1 per cent reduction in the ECE 15 urban cycle fuel consumption.
(b) Switching off the engine after a fixed-duration wait when the gearbox is in neutral, followed by automatic restarting on touching the accelerator pedal.
(c) Cutting off the fuel supply on over-run above a certain engine speed.

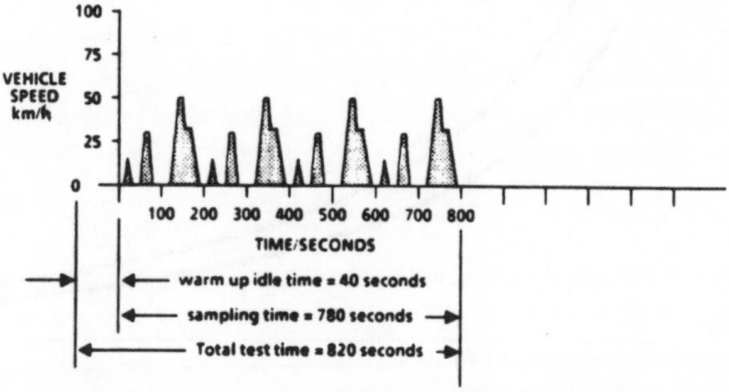

Figure 1.2 European ECE15 urban driving test cycle, from Barnes and Donohue (1985). [Reprinted with permission © 1985 Society of Automotive Engineers, Inc.]

The European (ECE) fuel consumption tests are conducted on vehicles which have been driven at least 3000 km. There are two constant speed tests at 120 and 90 km/h (75 and 56 mph), and the urban driving cycle; all the tests are conducted on fully warmed-up vehicles. The constant-speed tests can be conducted on test tracks, under strictly controlled weather and track conditions, or in a laboratory. The urban cycle has to be carried out in a laboratory using a chassis dynamometer (or rolling road). The driving wheels engage with rollers connected to a dynamometer which can simulate the air resistance; the inertia of the vehicle is modelled by flywheels with appropriate gearing. The rolling resistance of the tyres is different on the rollers from a flat surface; this is discussed further in section 6.2.2. It should be evident that great care is needed if there is to be reproducibility between test facilities. Furthermore, the vehicle has to be driven on its test pattern within specified tolerances. With a human driver there can be problems of repeatability, especially between different drivers, consequently robot drivers are also used. Advantage can, of course, be taken of the tolerance bands in the driving cycle to obtain optimum results.

The limitations of the fuel consumption tests include the use of fully warmed-up vehicles. This is particularly relevant to the urban fuel economy cycle. The fuel consumption penalties associated with short journeys using a cold engine are shown in figure 1.3 for two different ambient temperatures.

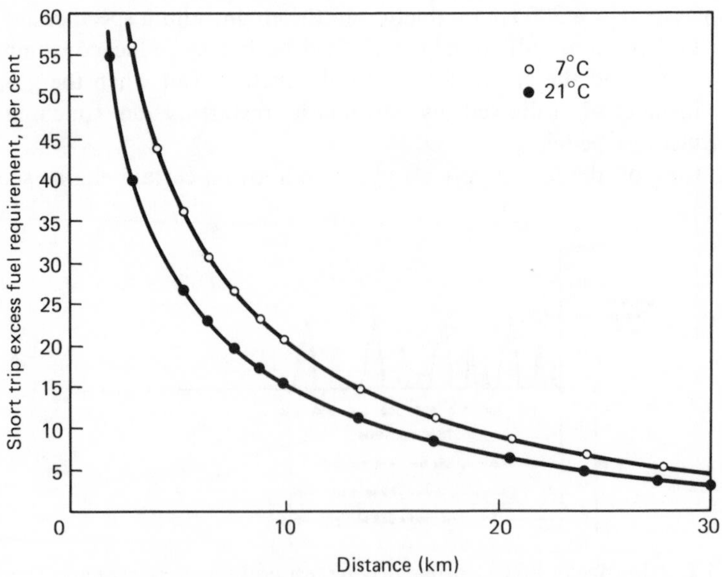

Figure 1.3 Excess fuel required for cold start trips, in a Ford Pinto at 72 km/h, adapted from Eccleston and Hurn (1979)

The excess fuel used is defined as the difference between that used in the test and that which would be used if the engine was fully warmed-up. The fuel consumption penalty needs to be examined in the context of trip length data, such as those shown in figure 1.4. The data in figure 1.4 can be manipulated to show the percentage of total miles as a function of trip length. When this is used in conjunction with figure 1.3, it is possible to calculate the percentage of excess fuel consumed because of having to warm-up the engine. If it is assumed that about half the total journeys commence with a cold start, then the total fuel consumed by a vehicle is increased by about 10 per cent through having to warm-up the engine. Studies conducted by the Transport and Road Research Laboratory (TRRL) have suggested that as much as 37 per cent of the petrol and 11 per cent of the diesel used in the UK is attributable to the excess associated with cold starting. [Pearce and Waters (1980)]. Improving the rate of engine warm-up would also lead to reduced emissions,

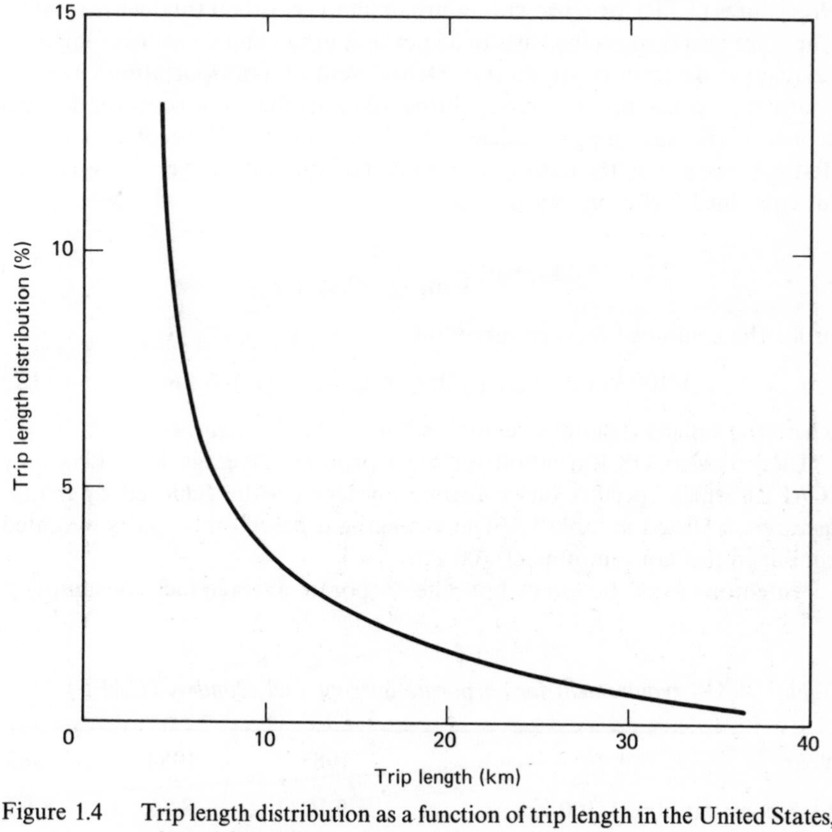

Figure 1.4 Trip length distribution as a function of trip length in the United States, adapted from Cole (1984)

less engine wear and lubricant degradation, and a better response from the interior heater. Both the US and European emissions test procedures start with a cold engine, and the US test also calculates the fuel consumption from the emissions test. However, the European urban fuel consumption is measured for a fully warmed-up vehicle.

During the cold starting of spark ignition engines, a very fuel rich mixture is used, to ensure that enough fuel evaporates to form a flammable mixture. This also leads to unburnt fuel forming unburnt hydrocarbons in the exhaust. In addition, with a cold combustion chamber, the extent of flame quenching will be greater, and this also leads to hydrocarbon emissions. Finally, the rich mixture also produces high emissions of carbon monoxide (the subjects of emissions and its legislation are discussed further in section 2.4.2). In the US emissions test, Kummer (1984) states that up to three-quarters of the permitted hydrocarbon emission can occur during the initial 120 seconds of the test; this is also a function of exhaust catalysts requiring time to reach their operating temperatures.

The US Environmental Protection Agency (EPA) has Federal Test Procedures (FTP) for representing urban and non-urban (highway) driving. These are averaged on the basis of 55 per cent urban and 45 per cent highway driving (as determined by the US Department of Transportation). It is, of course, fuel consumption (in, say, litres/100 km) that is of concern, not fuel economy (in, say, mpg)–vehicles are driven on the basis of arriving at destinations, not on the basis of using a certain quantity of fuel. The equation for combined fuel economy is thus

$$\text{mpg}_c = \frac{1}{0.55/\text{mpg}_u + 0.45/\text{mpg}_h} \tag{1.1}$$

or for the combined fuel consumption

$$(1/100\ \text{km})_c = 0.55\ (1/100\ \text{km})_u + 0.45\ (1/100\ \text{km})_h \tag{1.2}$$

where the suffices denote: c–combined, u–urban, h–highway.

There is also US legislation for the Corporate Average Fuel Economy (CAFE), which specifies fuel consumption levels to be achieved by manufacturers, as listed in table 1.4. The averaging is based on the sales weighted combined fuel consumption (1/100 km).

Knighton (1984) describes how the corporate average fuel consumption

Table 1.4 US requirement for corporate average fuel economy (CAFE)

Year	1983	1984	1985
Fuel consumption (1/100 km)	9.10	8.76	8.60

of Ford cars reduced from 16.66 l/100 km in 1974 to 9.94 l/100 km in 1983. However, the 1983 CAFE level was not met, and this was a result of free market demand–the fall in fuel prices in the 1980s produced a reversion to large engined cars.

In Europe there are no legal requirements for reducing the fuel consumption, nor is there are an agreed procedure for a weighted fuel consumption to represent an overall average fuel economy.

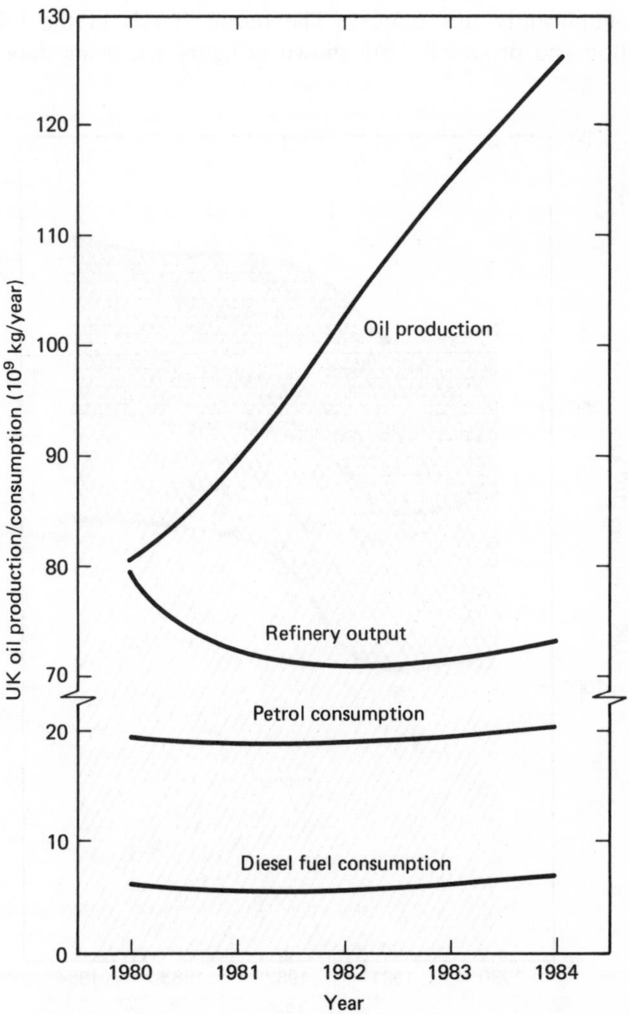

Figure 1.5 Recent trends in UK oil production and consumption; data derived from the Department of Energy (1985)

1.4 Fuel Resources

For as long as internal combustion engines have been used, there have been concerns about the future supplies of suitable fuels. However, the oil companies can only justify exploration costs to demonstrate reserves for a finite period. Consequently, the ratio of proven oil reserves to the rate of consumption is always about 20–30 years. New reserves are found in existing fields, and in addition new oil fields are found.

As a result of fuel conservation measures in all sectors, and the use of alternative fuel sources, the world consumption of crude oil is unlikely to increase significantly this century. The recent trends in the UK for oil consumption and production are shown in figure 1.5, using data from the

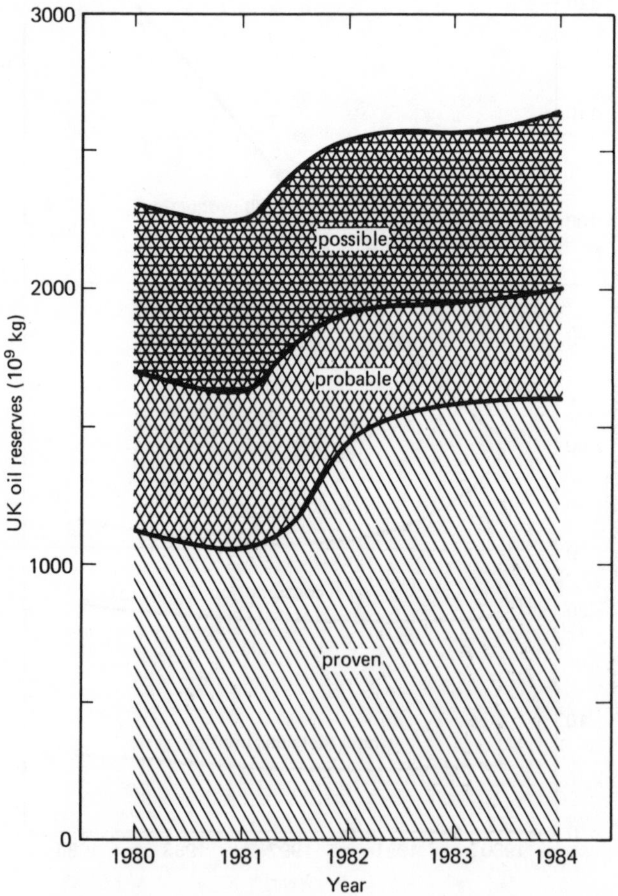

Figure 1.6 The changes in the UK oil reserves; data derived from the Department of Energy (1985)

Department of Energy (1981–5). It can be seen that petrol represents about 25 per cent of the refinery output, and that diesel fuel for road use represents about 8 per cent of the refinery output. After allowing for the oil used in the refinery (about 5 per cent of the throughput), figure 1.5 shows that the UK was self-sufficient in oil by 1981. For the same period, the UK oil reserves are shown in figure 1.6. The major UK oil field is in the North Sea, and this is now a well established field. None the less, discoveries continue to be made, and the proven reserves have increased, despite the increased production.

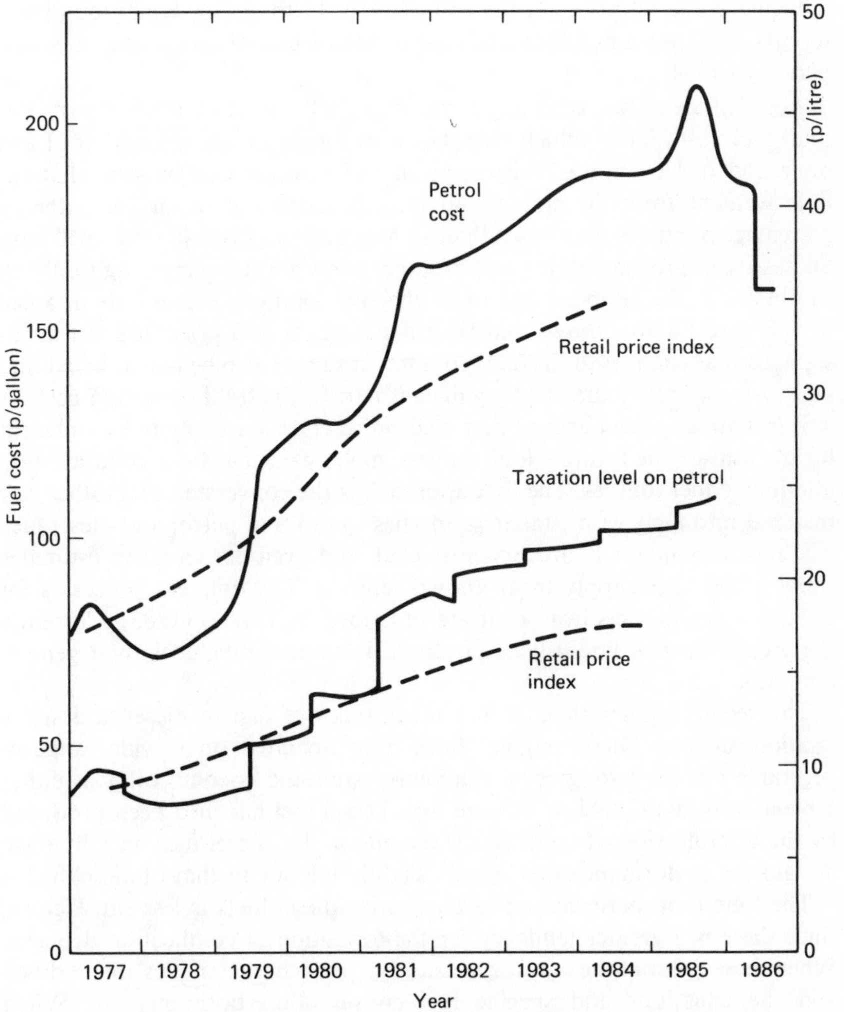

Figure 1.7 The changes in the retail price index, petrol cost and petrol taxation for the UK; data derived from the Department of Energy (1985)

The ratio of UK proven reserves to oil consumption is about 20 years. Since the Government policy is to export oil, production is greater than the domestic demand, and the ratio of reserves to production has fallen to about 15 years. When several governments hope to earn revenue from oil exports, there is liable to be a surplus of production over demand. This leads to a fall in fuel prices (in theory anyway), as shown in figure 1.7.

The cost of petrol and the element of taxation are both shown in figure 1.7, with the retail price index added for comparison. The level of taxation has risen faster than the retail price index, and this has resulted in the fuel cost rising faster than the rate of inflation, until the fall caused by reducing crude oil prices at the end of 1985. In the USA, Knighton (1984) reports a 23 per cent fall in the cost of petrol in real terms over a 3 year period to 1984.

The attitude to fuel economy is inevitably influenced by the fuel cost. The total fuel costs for a vehicle are shown in figure 1.8, as a function of fuel price and fuel economy. With a given fuel price, it can be seen that the improvement from 30 mpg to 40 mpg is more significant (on either a percentage or an absolute basis) than an improvement from 40 mpg to 50 mpg. Successive improvements in vehicle fuel economy are also increasingly difficult to obtain, as the simplest and most effective solutions are usually adopted first. Figure 1.8 also shows that typical changes in fuel prices are at least as significant as changes in the fuel economy. It should also be remembered that fuel costs are rarely the most significant part of the total operating cost.

It has already been argued here that oil reserves are likely to be sufficient for the foreseeable future. None the less, much work has been conducted on alternative fuel sources. The first approach is the conversion of another raw material into fuels with similar properties to those of petrol and diesel fuel. The most abundant hydrocarbon is coal, and even conservative estimates show a 200 year supply from known reserves. The different processes for obtaining liquid fuels from coal are described by Davies (1983). Currently the greater cost of liquid fuels produced from coal precludes their general adoption.

The second approach is to use other types of fuel in diesel and spark ignition engines. Diesel engines have been operated on a wide range of vegetable oils (including corn, sunflower, palm and coconut oils), as either a blend with diesel fuel or as pure fuel. Diesel fuel has also been produced by the esterification of lamb fat. These alternative diesel fuels usually have an ignition performance that is only slightly inferior to that of diesel fuel.

The long-term performance of alternative diesel fuels is less satisfactory, since there is a greater tendency for the formation of combustion deposits. When these occur at the injector nozzle, the spray characteristics are modified and the emissions and specific fuel consumption both increase. When combustion deposits occur in the piston ring grooves, the sealing and lubrication of the piston rings quickly deteriorates. Vegetable oils also degrade

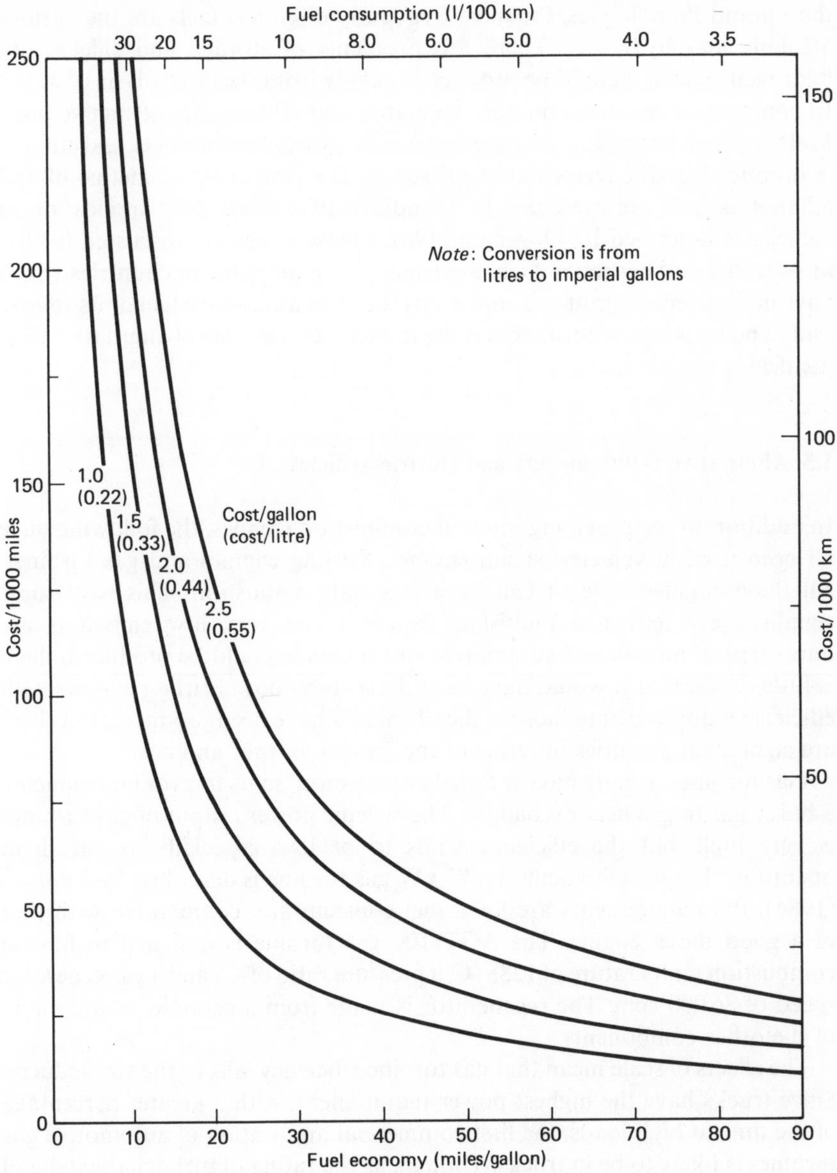

Figure 1.8 The effects of fuel price and fuel economy on fuel costs

the lubricant, and this too can lead to engine failure. Esterification can also be used to improve the properties of vegetable oils, and a very comprehensive summary has been produced by Onion and Bodo (1983).

The alternative fuels for spark ignition are usually more conventional. Apart from the gaseous hydrocarbons, which have very similar behaviour to their liquid homologues, the most frequently discussed fuels are the various alcohols and hydrogen. There are problems of storage and safety with hydrogen, even if it could be produced cheaply from the electrolysis of water. In contrast the alcohols, notably methanol and ethanol, are attractive fuels. Methanol can be produced by the destructive distillation of wood, and ethanol is produced by the fermentation of sugar. The properties of methanol and ethanol as fuels are presented by Goodger (1975), and their application to vehicles is described by Dorgham (1982); these issues are discussed further in section 2.5. Alternative fuels are usually only popular in countries which have no indigenous crude oil, and a surplus of an alternative fuel or its source. Only under these circumstances is the higher intrinsic cost of alternative fuels justified.

1.5 Alternative prime movers and electric vehicles

In addition to reciprocating internal combustion engines, the following have all been used in vehicles: steam engines, Stirling engines and gas turbines. All these engines have or can have external combustion. This is of some significance if either a multi-fuel capability or very low emissions are important. If an efficient automotive steam engine could be produced, there is little doubt that it would have been done so by now. Stirling engines with efficiencies approaching those of diesel engines have been produced, but there are significant penalties in terms of the specific output and cost.

Gas turbines usually have internal combustion, so as to avoid introducing a bulky gas to gas heat exchanger. The specific power output of gas turbines is very high, but the efficiency tends to be low–especially at part load operation. The development of a 73 kW gas turbine is described by Johnson (1984); the requirements are for a fuel consumption comparable with that of a good diesel engine. The AGT 100 gas turbine is designed to have a combustion temperature of 1288 °C, a pressure ratio of 4.5 and a gas generator speed of 86 000 rpm. The regenerator is made from a ceramic, as are many of the other components.

The effects of scale mean that gas turbine efficiency falls as the size reduces. Since trucks have the highest power requirement, with a greater percentage of the time at high loads, the first commercial application of automotive gas turbines is likely to be in trucks. Though, as the rating of turbocharged diesel engines increases, through the use of higher boost pressures, the thermodynamic differences between diesel engines and gas turbines are reducing. A quantitative comparison of different power plants is provided by Murphy (1985).

Electric vehicles offer an important alternative to engine-driven vehicles, but until recently their performance has been very restricted by the battery performance. Lead acid batteries have been developed over many decades, but 1 tonne of batteries only stores the same amount of energy as is contained in about 3.5 litres of fuel. Consequently, the performance of electric vehicles is restricted to a range of about 80 km, with a maximum speed of around 80 km/h. Data exist on the distribution of journey lengths, and in the USA for instance, 90 per cent of journeys are less than 65 km. However, the data that are really needed are the number of days in which the total distance travelled is less than the vehicle range. This is because the battery packs are usually recharged overnight.

In the UK, low-speed electric delivery vehicles have been very widely used, and considerable work has been invested in producing electric delivery vans with speeds of up to 80 km/h. Local delivery vans are an attractive proposition for electric propulsion, as they have a regular and restricted driving pattern. This maximises the battery life, and the regular use ensures maximum gains from the low running costs of electric vehicles. The developments that led to the introduction of the Bedford CF Electric delivery van have been described by Edwards (1984). The propulsive system was developed by Lucas Chloride Electric Vehicle Systems, and the main elements are the battery pack, controller, traction motor and the battery charger. Another advantage of a van is that the battery pack can be carried under the loading floor; in a car the battery pack invariably reduces the passenger and luggage space. In order to encourage the introduction of electric vehicles, the British Government provided a subsidy to offset the costs attributable to the small-scale production. The running costs of electric and spark ignition engine powered vehicles have been compared by Bedford (1985). The basis of comparison has been a daily distance of 72 km (45 miles) for 250 days a year. The results in figure 1.9 show that the costs per unit distance are comparable on the basis of an 8 year life. The comparison favours the electric vehicle if the spark ignition engined vehicle has only a 6 year life.

The capabilities of electric vehicles have now been significantly improved by the development of the sodium–sulphur battery [Scott (1986a)]. A cross-section of this battery is shown in figure 1.10, the sodium cathode (negative) and sulphur anode (positive) are both molten. During discharge the sodium ions migrate through the β-alumina insulator that separates the molten electrodes, to form sodium sulphide. The sodium sulphide also has to remain molten, so there is an operating temperature of 350°C; this temperature is maintained by the energy dissipated in the internal resistance. The sodium sulphur battery has an energy density of 520 kJ/kg, about four times that of a lead acid battery. Projections of vehicle performance with a sodium sulphur battery are for a range of 240 km, with a battery pack of 585 kg, half the weight of the usual lead acid battery pack.

Since this is a book on motor vehicle fuel economy, operating costs are not the only consideration and the overall energy efficiency of an electric vehicle also needs to be considered. In the first instance the electric power has to be generated, and with part load operation and the mix of power

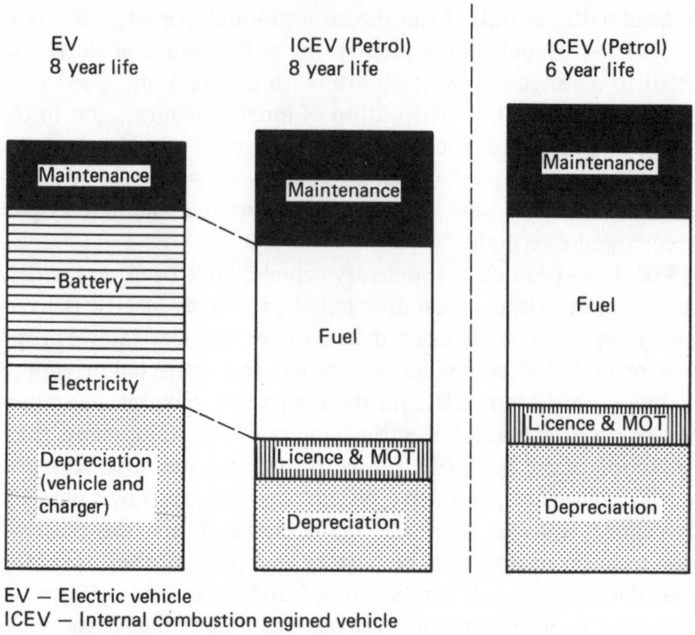

Figure 1.9 Comparison of the costs per unit distance for electric and spark ignition engine vehicles. [Reprinted with permission from Bedford (1985)]

stations the efficiency is about 30 per cent. Next the power has to be distributed, and with the various losses in power lines and transformers, the efficiency is about 90 per cent. With the battery and charger, only about 75 per cent of the electrical input to the charger is ultimately discharged by the battery. Finally, the electric motor and drive system will have an average combined efficiency of about 90 per cent. These efficiencies have been listed in table 1.5, to provide a simplified comparison with an internal combustion engined vehicle.

The comparisons in table 1.5 illustrate that the overall efficiencies of both approaches are comparable.

A wide-scale switch to electric vehicles is unlikely for two reasons. Firstly, the electric vehicle is only viable so long as its fuel is untaxed; if the taxation revenue from liquid fuels reduced, then governments would no doubt tax the electricity used in vehicles. Secondly, there is only a limited reserve generating capacity, even during the night. Thus electricity prices would rise if there was wide-scale use of electric vehicles.

Electric vehicles can make use of renewable and nuclear energy sources, but it has already been argued here that motor vehicle fuel supplies are sufficient (see section 1.4.). If concern increases about vehicle emissions, then

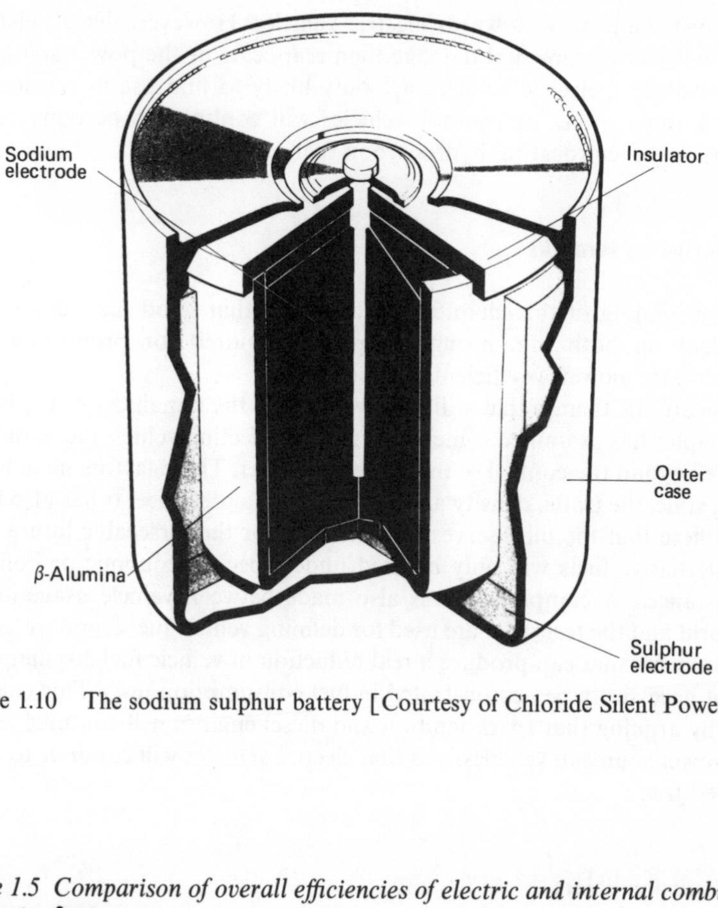

Figure 1.10 The sodium sulphur battery [Courtesy of Chloride Silent Power Ltd]

Table 1.5 Comparison of overall efficiencies of electric and internal combustion engined vehicles

Electric vehicle		Internal combustion engined vehicle	
System element	Efficiency (%)	System element	Efficiency (%)
Power station	30	Refining and distribution	90
Distribution	90		
Charger and battery	75	Engine at part load	20
Traction	90	Transmission	95
OVERALL	18	OVERALL	17

there might be a move towards electric vehicles. However, the problem of emissions or environmental damage then reappears at the power station.

In conclusion, electric vehicles are only likely to find use in a restricted range of applications, and motor vehicles will continue to be powered by conventional reciprocating hydrocarbon fuelled engines.

1.6 Concluding remarks

At the beginning of this chapter, it was stated that good fuel economy is dependent on both minimising the power required for propulsion and producing the power as efficiently as possible.

These are the themes that will be developed in the remainder of the book. This chapter has examined some of the factors affecting vehicle fuel economy that are beyond the control of the vehicle designer. These factors include the driving style, the traffic density and the vehicle maintenance. It has also been argued here that the oil reserves are adequate for the forseeable future, and that alternative fuels will only be used under special economic or political circumstances. A comparison was also made between vehicle usage in the real world and the tests that are used for defining vehicle fuel economy. Some improvements that can produce a real reduction in vehicle fuel consumption will not necessarily be demonstrated in fuel consumption tests. This chapter ended by arguing that spark ignition and diesel engines will continue as the main power source in vehicles, and that electric vehicles will continue to have a limited use.

1.7 Discussion points

(1) How do driving and driver characteristics affect motor vehicle fuel consumption?
(2) What influences driver attitudes towards fuel economy?
(3) How is vehicle fuel economy measured, and what are the limitations of test cycles?
(4) Why is vehicle fuel consumption worse in the real world compared with standard tests?
(5) What are the limitations of electric vehicles?
(6) Discuss the options and limitations of alternative fuels and prime movers.

2 Spark Ignition Engine Fuel Economy

2.1 Introduction

Both this chapter on spark ignition engines and the next chapter on diesel or compression ignition engines are intended to be self-contained. However, it is neither appropriate nor feasible to give full explanations of every aspect. Such information can be found in many texts, recent examples being Stone (1985) and Ferguson (1986). The differences between spark ignition and compression ignition engines arise from the means of ignition and mixture preparation. In spark ignition engines the mixture of air and fuel is prepared prior to induction, and this leads to pre-mixed combustion in which a flame front propagates across the combustion chamber. In contrast, the fuel is injected into diesel engines immediately prior to combustion. Time is required for the fuel to ignite and then burn in a mode known as diffusion combustion. The time required for the mixing, ignition and diffusion processes limits both the diesel engine speed and its air utilisation.

Spark ignition (SI) engines have very often been designed for high power outputs, and this is not always consistent with good fuel economy. The power output of an engine is a function of the flow rate of air and fuel through the engine, and the conversion efficiency from the chemical energy (of the fuel) to work. The fuel flow rate can be increased quite readily, but if effective combustion is to be maintained, then the air flow rate also has to be increased. For an engine with a fixed swept volume, the air mass flow rate can be increased either by pressurising the air supply, or by increasing the engine speed. A pressurised air supply usually requires a reduced compression ratio, and this reduces the efficiency, for reasons that are discussed in the next section. The turbocharging of both spark ignition and diesel engines is described in section 3.4.1.

When the output of an engine is increased by extending the speed range, the efficiency falls for two reasons. Firstly, as the speed is increased the power lost to friction increases at a greater rate than the gain in power; in other words, the mechanical efficiency falls. Secondly, if the volumetric efficiency (or breathing ability) of the engine is to be maintained at high speeds, then the value timing is characterised by large opening periods. For example, the period of valve overlap (during which both the inlet and exhaust valves are open), may be increased from 10° to 40° of crankshaft angle, in order to benefit from the dynamic effects in the induction and exhaust manifolds. However, when such an engine is operated at part throttle or a lower speed, then the dynamic effects are lost, and mixing occurs between the burnt and unburnt gases. In particular, unburnt fuel and air can pass straight into the exhaust system, and this obviously leads to a loss of efficiency.

Even when an engine has been designed for optimum fuel economy at full throttle, it will not necessarily lead to the optimum fuel economy in a vehicle. In section 4.2 it will be shown that a vehicle with a conventional transmission does not normally operate at the maximum engine efficiency condition. Thus, a better vehicle fuel consumption can be obtained by an engine with a lower maximum efficiency if the engine conditions where this occurs are closer to the vehicle operating conditions.

The parameters that have the greatest influence on engine efficiency are the compression ratio and the air/fuel ratio; both of these are discussed in section 2.2. In general, raising the compression ratio improves the efficiency, but at the expense of needing a higher-quality fuel (that is, of higher octane rating). The quality of the fuel can be improved by additives such as tetramethyl or tetraethyl lead, but environmental pressures are increasing the amount of legislation that limits the use of these materials. Alcohols have attractive properties as fuels, and they can also be used to extend and improve the quality of petroleum-derived fuels. Both these aspects are described in section 2.5. The quality of the fuel can also be improved by additional oil refinery processes. However, crude oil from various sources has quite different characteristics, and each refinery is designed for producing a different balance of fuels from a given input. Consequently, higher-quality fuels can only be produced at a greater expense, and this expense is both in the capital equipment and the energy used during the refining process. Thus the vehicles and refineries need to be treated as a single thermodynamic and economic system, and since the vehicle populations, legislation, crude oil and refineries all vary, there is not going to be a single best solution for the engine and oil refinery designs. Furthermore, even when there is a change in practice caused by legislation, the existing vehicles still have to be supplied with fuel.

The area in which engine designs differ most fundamentally is in the combustion chamber design, yet all are aiming to achieve a low fuel consumption with acceptable emisions. In section 2.4 several different combustion chamber designs are discussed, and these include lean burn engines, the Nissan NAPS-Z system and four-valve heads. Emissions legislation can have a significant influence on the combustion chamber selection, and any discussion also needs to include exhaust catalyst systems.

Areas that can easily be overlooked are the mechanical efficiency of an engine and the power requirements of any ancillary devices. The most obvious benefit of improving the mechanical efficiency or reducing ancillary power requirements is the increase in power that is available for traction. If the net power output is increased by 5 per cent, then since the fuelling rate will be the same, the efficiency would also be improved by about 5 per cent. However, the most significant improvements in efficiency will occur at part load operation, which of course is of much greater significance for a vehicle. In crude terms, the same 5 per cent reduction in parasitic power consumption would lead to about a 10 per cent improvement in fuel consumption at half load, and about 20 per cent at quarter load. Power consumption is incurred by the following items, typically in this order of significance: the air conditioning compressor, cooling fan, alternator, power steering pump and water pump. For good vehicle fuel economy, the power consumed by these ancillaries and their drives needs to be minimised. These factors affecting the mechanical efficiency are discussed in section 2.3; they are, of course, equally applicable to diesel engines.

Finally, if the optimum fuel economy is to be obtained, the engine and its cooling system need to be designed as a single system. If the engine and water pump are designed by an engine development section, and the radiator and cooling fan are designed by a chassis development section, then the radiator and fan are likely to be designed for ease of manufacture or low cost, rather than as part of an optimised system.

2.2 Spark ignition engine fundamentals

2.2.1 Compression ratio

The simplest model for the processes that occur in the spark ignition engine is the Otto cycle, and this shows how the efficiency varies with compression ratio.

$$\eta_{\mathrm{Otto}} = 1 - \frac{1}{r_{\mathrm{v}}^{\gamma-1}} \tag{2.1}$$

where r_{v} = volumetric compression ratio and $\gamma =$ ratio of the specific heat capacity for air, $c_{\mathrm{p}}/c_{\mathrm{v}}$.

While this model shows the correct trend of efficiency–increasing with the compression ratio–it overestimates the efficiency by a factor of about 2. The main reasons for this error are:

(a) During the compression and expansion processes there are both mechanical friction and heat transfer.
(b) The combustion process does not occur instantaneously.
(c) The exhaust valve opens before the end of the expansion stroke.

(d) The gas is in fact either a fuel/air mixture or combustion products, and these behave quite differently from the perfect gas that is assumed for the Otto cycle.

The major source of error in the Otto cycle model is the non-ideal behaviour of the air/fuel mixture and the combustion products; indeed, the behaviour depends on the air/fuel ratio. The strength of the air/fuel ratio can be described by the equivalence ratio, ϕ:

$$\phi = \frac{\text{Stoichiometric gravimetric air/fuel ratio}}{\text{Actual gravimetric air/fuel ratio}}$$

The stoichiometric ratio is the chemically correct ratio for complete combustion. If the mixture is rich, the equivalence ratio is greater than unity ($\phi > 1$), while if the mixture is weak or lean the equivalence ratio is less than unity ($\phi < 1$). The fuel/air cycle is a development of the Otto cycle that allows for the real thermodynamic behaviour of the gases, and the efficiency is shown for a range of compression ratios and equivalence ratios in figure 2.1.

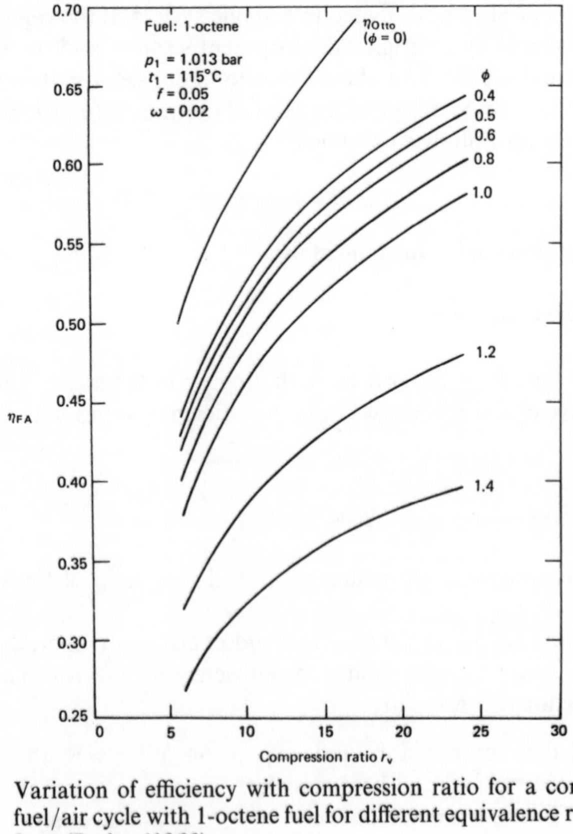

Figure 2.1 Variation of efficiency with compression ratio for a constant-volume fuel/air cycle with 1-octene fuel for different equivalence ratios, adapted from Taylor (1966)

Examination of figure 2.1 shows that as the mixture is weakened ($\phi \rightarrow 0$), the efficiency at a given compression ratio improves. In other words, the properties of weak mixtures are closer to those of a perfect gas that are assumed in the Otto cycle analysis. This is one of the reasons why the maximum efficiency of an engine occurs with a mixture that is weak of stoichiometric.

As the compression ratio is increased, the pressure and temperature of all the processes in the cylinder are raised. The highest pressures and temperatures occur during combustion, as the flame propagates across the combustion chamber. If the pressure and temperature of the unburnt mixture are high enough for a sufficient time, then the unburnt mixture will ignite spontaneously. Since the whole of the unburnt mixture can ignite spontaneously, the rate of pressure rise is much greater, and this produces a characteristic 'knock'; this self-ignition is also referred to as detonation and knock. The rapid pressure rise destroys the thermal boundary layer, and this can lead to overheating of components such as the piston and exhaust valve. Furthermore, hot surfaces (notably the exhaust value or combustion deposits) can act as sources of pre-ignition, that is, when ignition occurs before the spark. Under these circumstances the power output reduces, since the ignition occurs too early in the cycle. Also, the higher pressures and temperatures produced by pre-ignition can lead to self-ignition, if this has not already occurred. In other words, knock or self-ignition can lead to pre-ignition and vice versa. Either form of abnormal combustion can lead to a reduction in output and component failures.

The quality of a fuel is expressed by its octane rating, an indication of the resistance of a fuel to self-ignition. The octane rating is determined under carefully controlled conditions in an engine test. In an engine, the occurrence of self-ignition will depend on many parameters, including the compression ratio, air/fuel ratio, ignition timing, throttle setting, speed, air inlet temperature and combustion chamber design. The effect of raising the compression ratio is to increase the power output, and to reduce the fuel consumption; this is shown in figure 2.2.

The specific fuel consumption is proportional to the reciprocal of efficiency, and is more convenient for several reasons, including:

(a) The value is computed directly from the fuel flow rate and power output.
(b) If the power requirement is known for a particular vehicle as a function of its speed, then the fuel consumption of the vehicle can be found directly.

However, the specific fuel consumption does not take into account the calorific value of the fuel, and this is about 5 per cent greater for diesel than petrol. Often it is more convenient to use units of kg/kWh rather than kg/MJ, but the conversion is trivial (1 kWh $\equiv$ 3.6 MJ). The octane number requirements for a given compression ratio vary widely, but typically a compression ratio

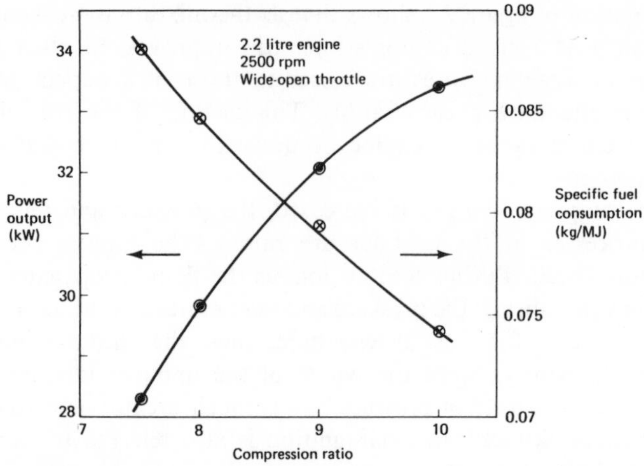

Figure 2.2 Effect of changing compression ratio on engine power output and fuel economy [with acknowledgement to Blackmore and Thomas (1977)]

of 7.5 would require 85 octane fuel, while a compression ratio of 10 might require 100 octane fuel.

From the foregoing it may appear that if an appropriate fuel were available, then much higher compression ratios would be beneficial; this is not in fact the case. As the compression ratio is raised, the incremental improvements in cycle efficiency reduce. Since the pressure loadings increase with the compression ratio, the frictional losses will also increase (for example, on the piston, the piston rings and the associated bearings), and this leads to a reduction in the mechanical efficiency. These trends are illustrated by figure 2.3, which suggests that the optimum compression ratio is about 14, and that there is not much to be gained from a compression ratio greater than about 10. Since mechanical efficiency is also dependent on speed, the optimum compression ratio for a given engine will vary with speed. The values quoted for the optimum compression ratio will depend on the engine and operating conditions, but the values are usually in the range of 14–17.

2.2.2 Air/fuel ratio

The mixture strength or air/fuel ratio affects the efficiency of the spark ignition engine in several ways. It has already been shown that the air/fuel ratio influences the efficiency of the corresponding cycle (figure 2.1). The richer the mixture, the less perfect the behaviour of the gas and the lower the efficiency.

The mixture strength also affects the combustion process; indeed if combustion is to occur at all, the mixture has to be close to the stoichiometric

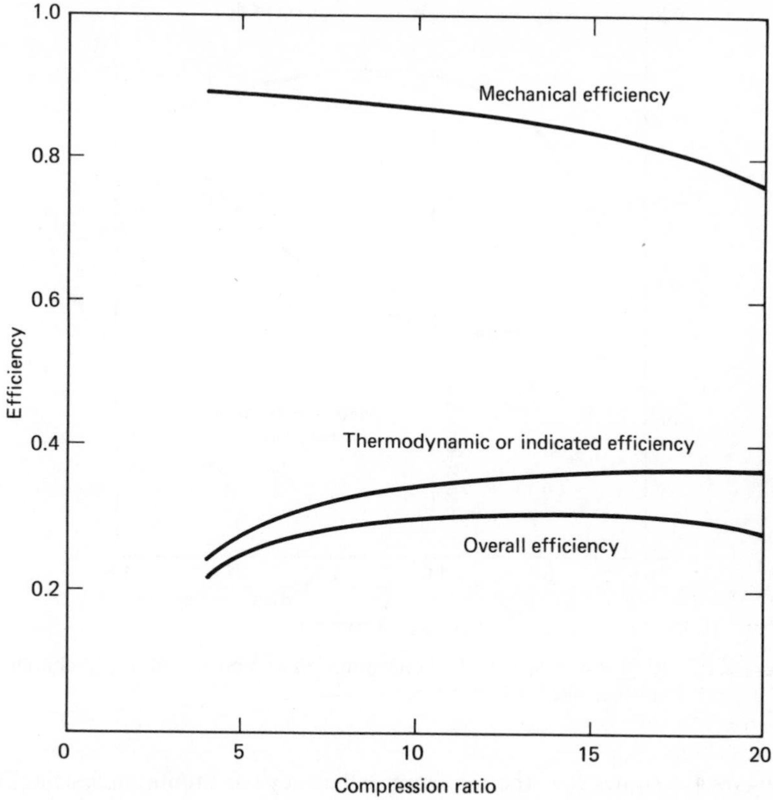

Figure 2.3 The effect of compression ratio on efficiency for a given engine and operating condition

value. The responses of the specific fuel consumption and power output to changes in the air/fuel ratio are shown in figure 2.4, for an engine at a fixed speed and throttle setting. The power output is plotted in terms of the brake mean effective pressure (bmep), since this gives an indication of the engine output, in a form that is independent of the swept volume and speed of an engine.

$$\text{Power,} \quad \dot{W} = p_b L A N' \tag{2.2}$$

where

p_b = bmep

L = engine stroke

A = total piston area

and

N' = number of firing strokes/second/cylinder.

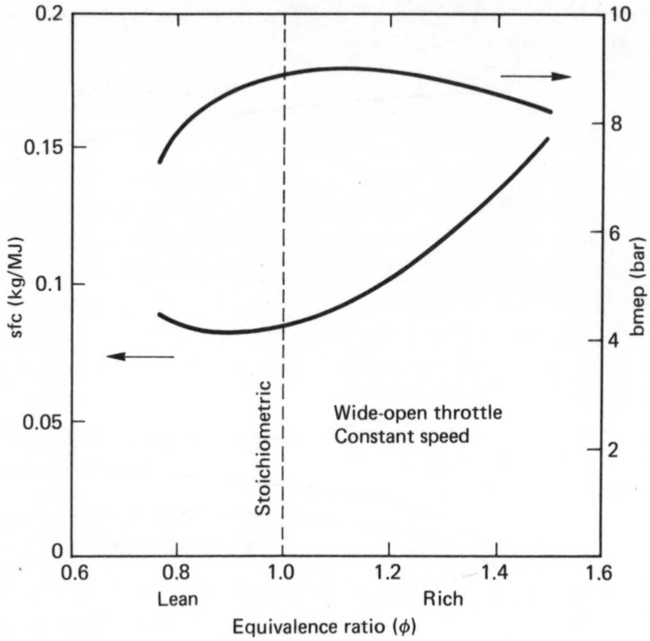

Figure 2.4 Response of specific fuel consumption and power output to changes in air/fuel ratio

Figure 2.4 shows that the maximum efficiency (or minimum specific fuel consumption) occurs with a mixture that is weaker than stoichiometric. In contrast, the maximum power output occurs with a mixture that is richer than stoichiometric. The reason that both the maxima do not occur with a stoichiometric mixture is mainly attributable to the occurrence of dissociation. At the high temperatures that occur with the combustion process, an equilibrium could be reached between the reactants and products. As an example, carbon monoxide (CO), oxygen (O_2) and carbon dioxide (CO_2) can all co-exist in equilibrium:

$$CO + \tfrac{1}{2}O_2 \rightleftharpoons CO_2$$

The position of the equilibrium is influenced by the pressure, temperature and composition of the mixture. If the mixture is rich, then the excess fuel leads to partially burnt fuel, and the high level of carbon monoxide moves the equilibrium to the right, thus maximising the oxygen utilisation and the power output. Too much fuel, though, will further reduce the cycle efficiency and increase the amount of partially unburnt fuel.

Conversely, with a lean mixture, the greater relative amount of oxygen will help to maximise the fuel utilisation. While a leaner mixture increases the cycle efficiency, a point will be reached at which the combustion of the mixture becomes unreliable. However, before this point is reached the maximum overall efficiency will occur, because when the power output is reduced the mechanical losses become more significant and the mechanical efficiency reduces. This is because the overall efficiency is the product of the cycle efficiency and the mechanical efficiency.

The ignition limit of a weak mixture will depend on the local geometry, temperature and pressure. A high-energy ignition source can cause a weaker mixture to start burning, but it will not be self-sustaining. A small level of turbulence can also cause a weaker mixture to burn, but more significantly for engine combustion, a high level of turbulence richens the weak mixture limit. In other words, it is possible for a mixture to be ignited, but then to be extinguished as a consequence of a high turbulence level. Daneshyar *et al.* (1983) discuss the effect of turbulence in terms of the strain-rate on flame fronts, and they argue that weak mixtures are particularly sensitive to straining. Furthermore, strain fields reduce the burning velocity, and can extinguish flames. Daneshyar *et al.* argue that the effects of strain will be greatest in the early stages of combustion, and that this might be a significant contribution to the cycle to cycle variations in combustion.

The data from figure 2.4 can also be cross-plotted, and this can be seen in figure 2.5, along with some additional part throttle data. Points A and B on figure 2.5 represent the same power output, but with different specific fuel

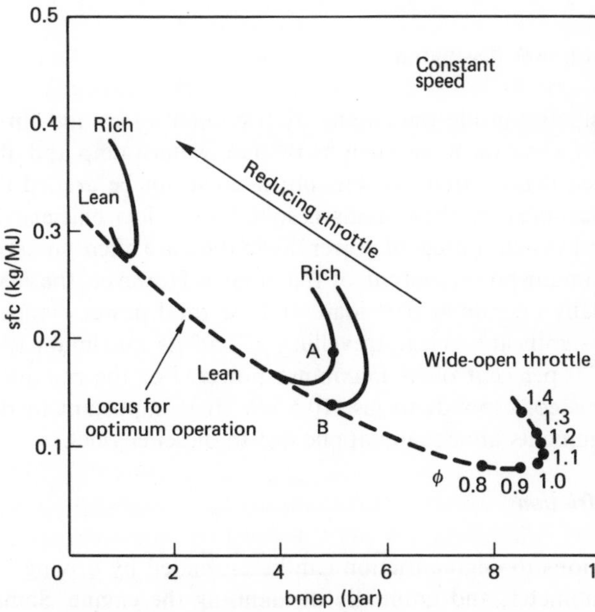

Figure 2.5 Specific fuel consumption plotted against power output for varying air/fuel ratios at different throttle settings

consumptions. The lower fuel consumption is obtained with a slightly wider open throttle and a weaker mixture. Any particular output (below the maximum) can be obtained from a range of throttle and mixture strength combinations, and the locus for optimum operation is also shown in figure 2.5. If the results shown in figure 2.5 are obtained for a range of speeds, then the resulting loci for optimum operation can be combined to produce a fuel consumption map; an example of this is figure 4.1.

Regardless of whether the air/fuel mixture is prepared by an injection system or carburettor, the requirements are essentially the same. At full throttle a rich mixture is needed to obtain maximum power output, but at part throttle settings a lean mixture is required to obtain the optimum fuel economy. It goes without saying that in a multi-cylinder engine the air/fuel ratio should be the same for all cylinders. With a multi-point fuel injection system this is quite simple, since there is one injector associated with each cylinder. When a carburettor or single point fuel injection system is used, then the mixture distribution is very sensitive to the inlet manifold design. When the mixture is prepared, the fuel is initially in the liquid state and the evaporation requires a finite time. The fuel flowing into the engine arrives as vapour mixed with air, as droplets, and as a liquid film flowing along the walls of the inlet manifold. Despite the careful attention paid to manifold design, it is quite usual to have a ± 5 per cent variation in mixture strength between cylinders, even for steady-state operation.

2.3 Sources of power dissipation

Power is dissipated inside the engine by frictional losses and through the inefficiencies of ancillary items such as the fan, water pump and alternator. Obviously such items require power, but it must not be argued that since the power consumed by these items is low, then a low efficiency will not matter. The individual sources of power dissipation are often small compared with the maximum power output of the engine. However, the majority of engines normally operate at part load, and the total power dissipation can thus be very significant. A car travelling at half its maximum speed may require only 20 per cent of its maximum power. For the car discussed in section 4.2, this corresponds to just 10.5 kW. It is important to remember that these arguments are equally applicable to diesel engines.

2.3.1 Engine friction

The contributions to engine friction can be estimated by driving an engine with a dynamometer, and gradually dismantling the engine. Some typical results for a range of speeds can be seen in figure 2.6. However, it must be remembered that in this type of test the pressure loadings from combustion are not present, and the frictional losses in some components will be undeterestimated. This is particularly so for the piston ring friction, but less so for the piston skirt, connecting rod and crankshaft bearings.

Measuring the frictional losses in a firing engine is difficult, since it depends on knowing accurately the indicated power. The indicated power is the rate at which net work is done by the gases on the piston, and this depends on measuring the pressure as a function of engine rotation. The difficulties arise for several reasons:

(a) The pressure varies over a wide range.
(b) Wide temperature variations during each cycle might affect the pressure transducer calibration.
(c) The position of top dead centre has to be known accurately ($<1°$).
(d) The indicated work is the difference between the compression and expansion work, minus the work dissipated in the gas exchange process.

Finally, since the friction work is the difference between the indicated work and the brake work (engine output), and these are comparable in magnitude,

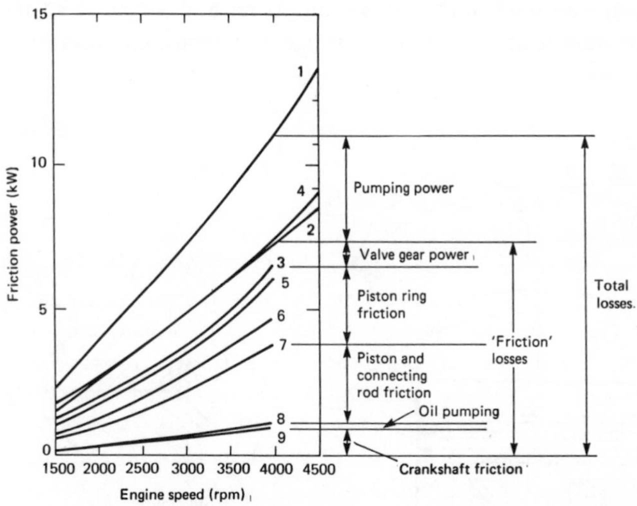

Figure 2.6 Analysis of engine power loss for a 1.5 litre engine with oil viscosity of SAE 30 and jacket water temperature of 80°C. Curve 1, complete engine; curve 2, complete engine with push rods removed; curve 3, cylinder head raised with push rods removed; curve 4, as for curve 3 but with push rods in operation; curve 5, as for curve 3 but with top piston rings also removed; curve 6, as for curve 5 but with second piston rings also removed; curve 7, as for curve 6 but with oil control ring also removed; curve 8, engine as for curve 3 but with all pistons and connecting-rods removed; curve 9, crankshaft only [adapted from Blackmore and Thomas (1977)]

any error in the indicated work produces a much larger percentage error in the estimate of friction work.

Thus the benefits of reduced friction (which, of course, have the most relevance) are often quoted, rather than the reduction in friction. However, Baker (1984) quotes some Volvo results, in which a 4 per cent economy improvement at idle is attributed to a 10–15 per cent reduction in friction. Friction is reduced by careful attention to balancing, bearing design and the piston. Despite the results in figure 2.6 only referring to an unfired engine, the trends are still indications of what might be expected in a fired engine. Since the friction power rises faster than a linear increase with speed, the mechanical efficiency will be significantly lower at high speeds, and the advantages of restricting the speed range of an engine should be self-evident.

A significant recent development is the AE (Automotive Engineering) Group AEconoguide piston. Viscous friction originates in the oil film between the piston skirt and the cylinder wall. To reduce this friction the bearing area is reduced, by machining the piston skirt to leave raised bearing pads, as shown in figure 2.7. The AEconoguide piston and its performance are described very thoroughly by Rhodes and Parker (1984). Engine tests suggest that there is a 10 per cent reduction in friction, and this leads to about a 2 per cent improvement in the maximum power and a 3 per cent reduction in the fuel consumption over a wider range of operating conditions.

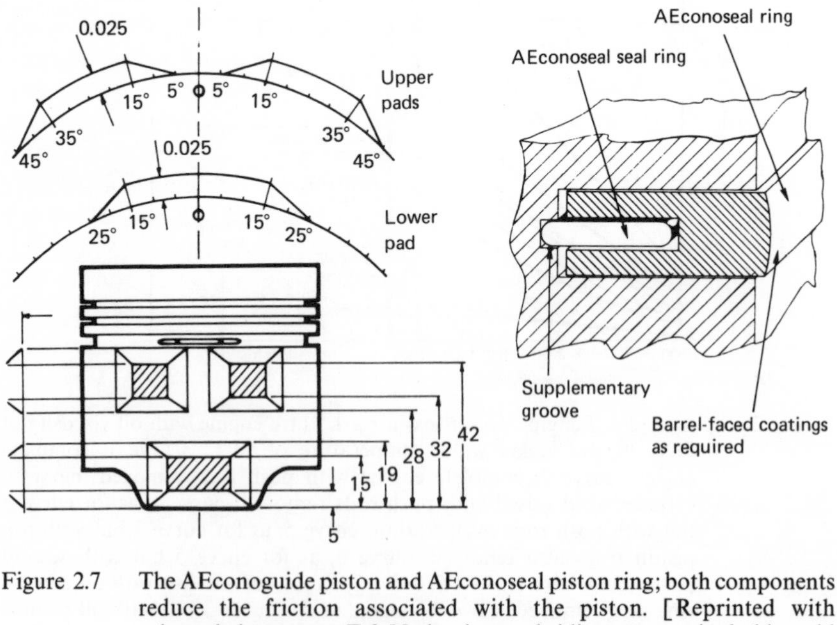

Figure 2.7 The AEconoguide piston and AEconoseal piston ring; both components reduce the friction associated with the piston. [Reprinted with acknowledgement to T & N plc whose subsidiary companies hold world patents on these designs]

Another development that is designed to reduce piston friction is the AEconoseal; this is also illustrated in figure 2.7. The AEconoseal is designed to replace both the upper and middle piston rings of a conventional piston. The reduced piston ring loadings reduce the friction, and eliminating one ring allows the piston to be shallower with a 10 per cent weight reduction.

Friction associated with other components in fired engines can be investigated in specially adapted engines. For example, the camshaft torque can be found from a strain-gauged camshaft drive wheel. Alternative techniques for finding the piston friction are to strain-gauge the connecting rod, and to use load cells on a floating cylinder liner.

The choice of lubricant will have a direct effect on the engine efficiency. However, the scope for improving engine efficiency by reducing the oil viscosity is limited, since low lubricant viscosity can lead to problems such as engine wear and high oil consumption. Numerous tests have been conducted to investigate the fuel economy benefits of different lubricants, but the comparison of results from different tests is impossible because oil viscosity (which is considered to be the variable in the experiments) is not measured under the conditions of high shear that occur in engines. The viscosity of oils is maintained at high temperatures by polymeric additives (viscosity index, VI improvers). Under conditions of high shear, the thickening effects of these additives are reduced to a lesser or greater extent. Thus the simple oil specifications, which are all that are usually quoted, are inadequate for comparisons of engine or vehicle fuel economy.

That oils with reduced viscosity improve vehicle fuel consumption is not in doubt. Furthermore, such improvements are greatest in cold start tests, since these conditions highlight the differences in viscosity. Unfortunately, cold start tests are also the most difficult to define and quantify. Results from such a test are reported by Toft (1984) for eight different vehicles in which a low viscosity oil (15W/40) was compared with a reference oil (20W/50). The mean fuel economy improvements varied between 6.0 per cent in the first 1.6 km to 1.9 per cent when fully warmed up.

The results from another series of tests are reported by Goodwin and Haviland (1978). In general, the largest improvements arose in urban driving with a 2–4 per cent reduction in fuel consumption obtained by adopting a low viscosity oil (5W/20 as opposed to 10W/40). For highway driving, the same change in oil produced reductions in the fuel consumption from 0 to 2 per cent.

2.3.2 Ancillary items

Many ancillary items (for example, the power steering pump, the air conditioning compressor and, to a lesser extent, the alternator) have a very unpredictable duty cycle. Significant loads on the alternator are caused by the heated rear window (about 10 A), the headlights (10–20 A) and recharging the battery. Paradoxically, recharging the battery is most significant immediately after starting an engine, when the fuel economy of the engine is worst. The maximum output of a typical automotive alternator is about 600 W; a notable exception is the alternator of a coach or bus that has to meet a large interior lighting load. The efficiency of an alternator at its rated load is shown in figure 2.8. The maximum efficiency is only 35.5. per cent and the efficiency can be expected to fall at part load.

The power consumed by the cooling system of an engine is much easier to define, since it is solely a function of the engine speed and the ambient conditions. Whether an engine is directly air-cooled or indirectly cooled with water as an intermediate heat transfer medium, the power used to provide the air flow can be a major consideration. The efficiency can be as low as 10 per cent for a pressed steel fan, while carefully designed fans made from injection moulded plastic can have efficiencies of over 30 per cent. For a water-cooled engine, the fan can absorb as much as 5 per cent of the power at the rated speed and load. The effect of improving the fan efficiency is illustrated by the example given in section 2.7.

The power absorbed by the fan can be reduced by using a viscous coupling, which has increasing slip with rising speed. An alternative approach is to have a thermostatically controlled fan that only provides a drive when the air temperature at the fan hub exceeds a pre-determined value. A more radical approach is to have a fan driven by an electric motor that is only operated when the temperature of the water leaving the radiator rises above a fixed

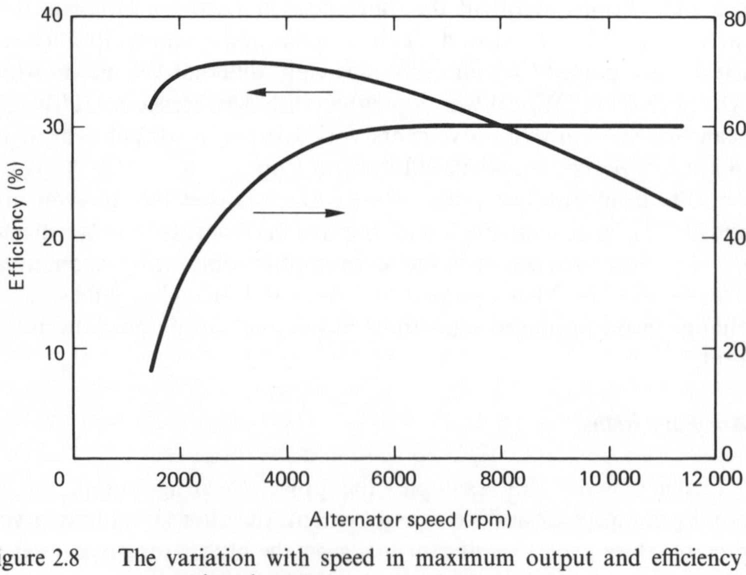

Figure 2.8 The variation with speed in maximum output and efficiency for an automotive alternator

value. This system is now common on cars, and the fan will often only operate in urban driving or after high-speed driving. The gains when the fan is not running have to be offset against the inefficiencies of the alternator and motor when the fan is running.

The maximum power requirement for the water pump is usually less than a couple of per cent of the rated engine output. However, significant improvements in efficiency are associated with the careful design of plastic injection moulded impellers. Finally, the efficiency of the ancillary drives must not be forgotten. In the past the common practice has been to use Vee belts, but the maximum efficiency of these is typically only 85–90 per cent. Toothed belts and Poly vee belts have efficiencies as high as 98 per cent.

2.4 Efficient spark ignition engines

The importance of a high mechanical efficiency has already been discussed; the other main requirements to produce an efficient engine are to have effective gas exchange processes, mixture preparation, ignition and efficient combustion. However, it is also essential to ensure that the engine will have sufficiently low emissions to satisfy any legislation, and be able to operate with the fuels that are likely to be available in the future.

2.4.1 Gas exchange processes

The valve timing and design of the induction and exhaust system have a profound effect on the shape of the torque *versus* speed curve. The maximum torque at any speed is essentially dependent on the amount of air/fuel mixture induced into a cylinder for combustion, and the effectiveness of removing all the combustion products from the previous cycle. The volumetric efficiency of an engine can be improved by tuning the induction and exhaust systems. This involves selecting the lengths and the geometry of the intersections in the induction and exhaust systems, in order to benefit from the dynamic flow effects. The dynamic flow effects are compression and expansion waves that are generated as a result of the intermittent nature of the flow to and from the engine cylinders.

If the effectiveness of the gas exchange process is improved, then this should manifest itself as an increase in the engine torque, with a corresponding increase in the power. If the engine was then reduced in size to maintain the same power level, then the associated friction should reduce and this will lead to an efficiency improvement at all the operating points.

The difficulty with designing the induction and exhaust systems is that the optimum lengths and valve timings will be a function of the engine speed. Since the valve timings and the geometry of the induction and exhaust systems are usually fixed, then great care is needed to ensure that an improvement at a particular speed does not lead to an overall deterioration in performance. Experimental variable geometry induction systems have been developed on a number of engines, including Diesel engines [Watson (1983)]. Yamaguchi (1986a) describes a system used by Honda on the C20A engine (the C25A derivative of this engine is used in the Rover 800, the case study in section 7.2). Honda use two intake tracts for each cylinder, a long (750 mm) tract of small diameter (27 mm) being used for low speeds. Above 3500 rpm, a solenoid-actuated and vacuum-operated butterfly valve selects a shorter tract (410 mm) with a larger diameter (37 mm). A slightly different system is used by Nissan, in which the inlet ducts for each bank of a V6 engine enter separate resonating volumes. Above 4400 rpm and 0.135 bar manifold vacuum the resonating volumes are connected together, to improve the high-speed air flow [Yamaguchi (1986c)].

Variable valve timing systems have been discussed for many years, and a review that summarises the benefits and classifies some of the different mechanisms has been presented by Stone and Kwan (1985). In addition to raising the torque curve, variable valve timing has been demonstrated to reduce emissions, and to improve the full and part load efficiency.

Variable valve timing has been most widely used to control the period of inlet and exhaust valve overlap around top dead centre. The first commercial application of this was by Alfa Romeo, and another recent example is from Nissan [Yamaguchi (1986c)]. The available valve timings are summarised in table 2.1.

This V6 engine has twin overhead camshafts on each cylinder bank, and the valve timing variation is obtained by adjusting the phasing of the inlet camshafts through 14°. The larger valve overlap (with inlet valve opening 19°btdc) is used at high loads, and speeds below 5200 rpm. Results that are typical of the improvement in full load performances are shown in figure 2.9.

A study of variable valve timing for spark ignition engined vehicles by Ma (1986) suggests that the urban cycle (ECE-15) fuel economy can be improved by as much as 15 per cent. Of this, 6 per cent would be attributable to a lower idling speed that reduces the idle fuel consumption. Variable valve

Table 2.1 Nissan variable valve timing performance

Valve event	Crank angle
Exhaust open (°bbdc)	55
Exhaust close (°atdc)	13
Inlet open (°btdc)	19 or 5
Inlet close (°abdc)	49 or 63

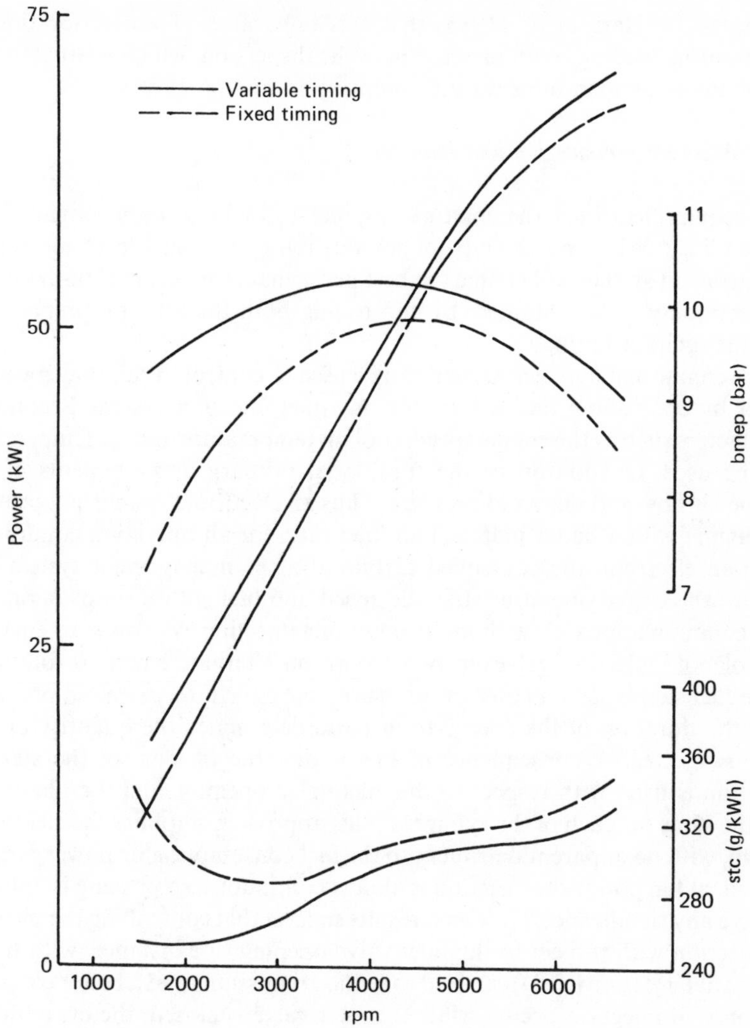

Figure 2.9 The effect of variable valve timing on spark ignition engine performance, adapted from Torazza (1972)

timing also offers scope for reducing the pumping work, which is dissipated by the pressure drop across the throttle at part load. At part load the mass of both fuel and air has to be reduced, and either early (before bdc) or later inlet valve closing are alternatives to throttling. Figure 2.10 shows the reason for the reduced pumping work with early inlet valve closing. Ma (1986) reports a 7 per cent reduction in fuel consumption at 30 per cent load by using early or late inlet valve closing. The reduction in fuel consumption is less than that suggested by the reduction in pumping work. It has been suggested by Hara *et al.* (1985) that this is because of a deterioration in combustion, leading to an increase in cyclic dispersion, which is attributable to the lower temperatures during compression and combustion.

2.4.2 Mixture preparation and ignition

The requirements for the mixture air/fuel ratio have been explained in section 2.2.2. Whatever the type of combustion system, an electronic engine management system will enable the best performance in terms of both output and economy to be obtained, by controlling both the mixture preparation and the ignition timing.

The engine management system can be used to control the air/fuel mixture, either by controlling the carburettor or a fuel injection system. Secondary parameters such as the engine speed, coolant temperature and air temperature can be used, in addition to the traditional primary measurements of the engine air flow and manifold pressure. Thus an electronic system can provide the engine with a better matched air/fuel ratio for all operating conditions. With an electronically controlled carburettor, the management system can control the choke operation, the idle speed and fuel cut-off on over-run.

A common approach with multi-point fuel injection systems is to energise the solenoid injection valves in two groups on alternate engine revolutions. If the fuel supply is at a constant pressure above the inlet manifold pressure, then the duration of the energisation pulse determines the quantity of fuel that is injected. A consequence of this is that the phasing of the start of injection is fixed with respect to the inlet valve opening, and the phasing is not the same for each of the cylinders. This approach simplifies the electronic design, with no apparent detriment to the fuel consumption or power output. Indeed, at full power the injection is almost continuous, so phasing is unlikely to have any significance. However, results suggest that controlling the phasing of injection with respect to the inlet valve opening (for instance, with timed sequential injection (TSI)) can lead to reduced emissions and a better transient response. If injection occurs while the inlet valve is closed, the evaporation from the back of the valve and the inlet port aids mixture preparation during steady-state operation, thereby leading to low emissions. Conversely, for transient operation, if the injection occurs while the inlet valve is open, the delays associated with the fuel film in the inlet port can be eliminated, and the transient response is quicker. The fuelling requirements are usually stored in a memory, with the values modified by the influence of secondary parameters such as the coolant temperature.

The ignition timings can also be stored in a memory and used to initiate a high-energy electronic ignition system. A map of the optimum ignition timings is stored as a function of the engine speed and manifold pressure. Consequently the compromises associated with limited controls, provided by the traditional centrifugal and vacuum advance systems, are eliminated.

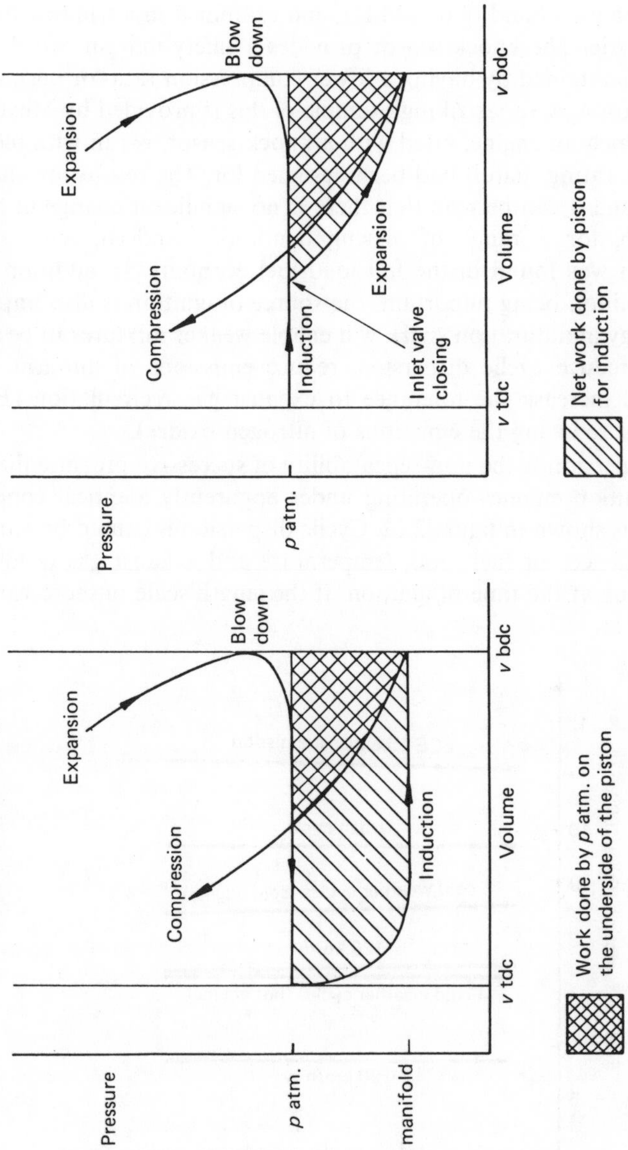

Figure 2.10 Early inlet valve closing as a means of reducing the pumping work, Stone and Kwan (1985). [Reprinted by permission of the Council of the Institution of Mechanical Engineers]

If a knock detector is fitted, this can retard the ignition at the onset of knock, thereby preventing damage to the engine. When an accelerometer is used as a knock sensor, Forlani and Ferranti (1987) report that the signal is typically filtered with a pass band of 6–10 kHz, and examined in a window from tdc to 70° after tdc. The knock sensor provides a safety margin, which would otherwise be obtained by having a lower compression ratio or permanently retarded ignition. A very striking example of this is provided by Meyer *et al.* (1984), in which an engine, fitted with a knock sensor, is run with fuels of a lower octane rating than it had been designed for. The results are shown in figure 2.11, and it can be seen that there is no significant change in the fuel consumption, for a range of driving conditions. Indeed, only a slight deterioration was found in the full load fuel economy. In addition to the timing of ignition being important, the source of ignition is also important. A high-energy and duration spark will enable weaker mixtures to be ignited and burnt, reduce cyclic dispersion, reduce emissions of unburnt hydrocarbons and increase the tolerance to exhaust gas recirculation (EGR–a technique for reducing the emissions of nitrogen oxides).

Cyclic dispersion is the non-repeatability of successive pressure diagrams, in spark ignition engines operating under apparently identical conditions; an example is shown in figure 2.12. Cyclic dispersion is caused by variations in the turbulence, air fuel ratio, temperature and exhaust gas residuals, at the spark plug at the time of ignition. If the length scale of these variations

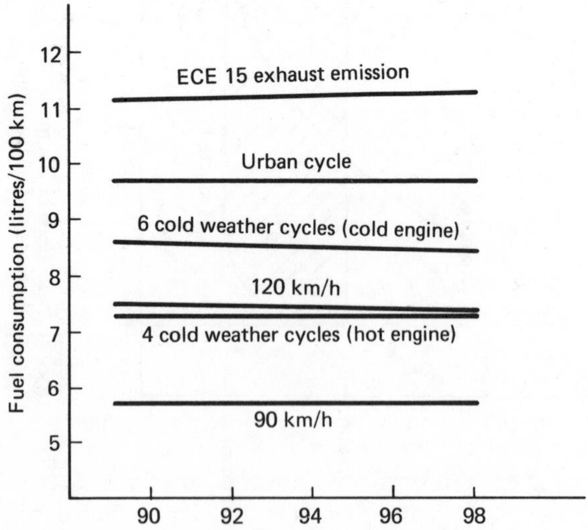

Figure 2.11 The effect on fuel consumption of reduced fuel quality on an engine designed for 98 octane fuel, but fitted with a knock sensor controlled electronic ignition, adapted from Meyer *et al.* (1984)

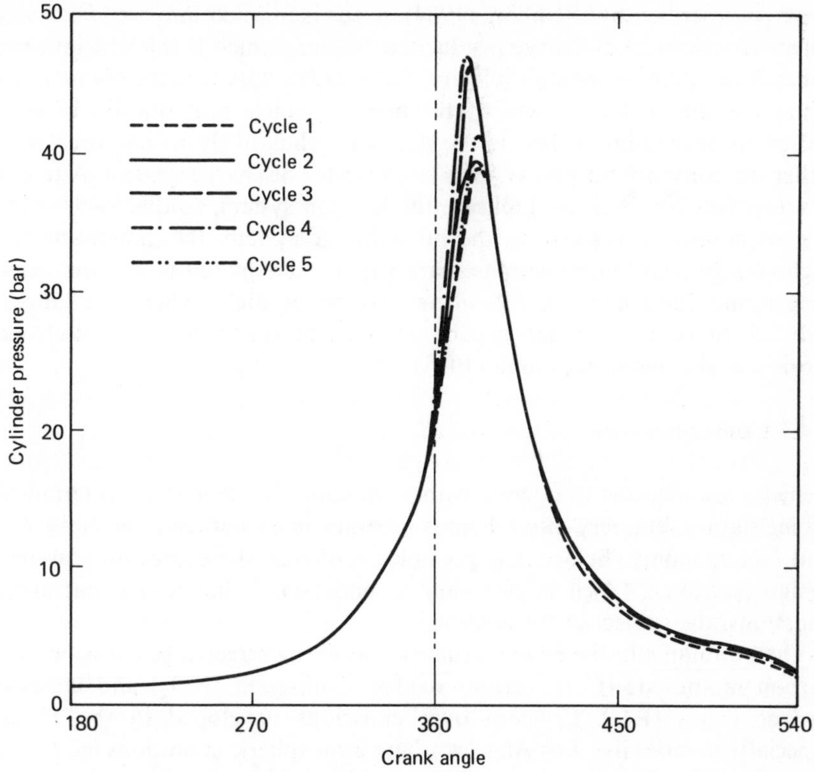

Figure 2.12 Pressure–time diagrams for five successive cycles in a Ricardo E6 engine (compression ratio 8:1, stoichiometric air/iso-octane mixture, 1000 rpm, 8.56 bar bmep), from Stone and Green-Armytage (1987). [Reprinted by permission of the Council of the Institution of Mechanical Engineers]

is greater than the size of the spark or flame nucleus, then the variations will affect the propagation of the flame front. Furthermore, the initial flame propagation affects the subsequent development; a charge that starts by burning slowly can be seen to continue to burn slowly. Cyclic dispersion is worst with lean mixtures at part load operation [Hancock *et al.* (1986)], and these are operating conditions that will become increasingly significant in automotive applications. Each of the pressure diagrams shown in figure 2.12 will give a different indicated power, and it is self-evident that only one can be the optimum. Thus if cyclic dispersion can be reduced, then the overall efficiency and the driveability will both be improved.

The physics of sparks and ignition is treated very comprehensively by Maly (1984), who also discusses the performance of many different systems in engines. A more radical approach is to use a plasma jet ignition system, of the type described by Weinberg (1983). A gas is injected into a small cavity where an electrical discharge produces a plasma, which is injected into the combustion chamber at high velocity. The disadvantages are the high-energy input and the erosion caused by the plasma, which limits the life of even pulsed plasma ignitors. Ignition systems are thus likely to continue with either the conventional coil systems or capacitor discharge systems. A recent development has been to duplicate the ignition system, so that each spark plug is connected directly to the high-voltage system. By eliminating the high-voltage distribution with its spark gap, the energy released at the spark plug is increased and the risk of an extraneous high-voltage breakdown reduced. Since a wider gap can be used at the spark plug, the electrode erosion is also reduced [Dunn (1985)].

2.4.3 Emissions

Vehicles are required to operate within emissions levels that are determined by legislation, but very often there is a compromise between low emissions and fuel economy. Before it is possible to discuss the combustion system design (section 2.4.4) it is necessary to understand the factors that affect emissions, the subject of this section.

The pollutants in the engine exhaust that are referred to as emissions are carbon momoxide (CO), various oxides of nitrogen (NO_x) and unburnt hydrocarbons (HC). Concern over emissions developed in the 1960s, especially in cities like Los Angeles where atmospheric conditions led to the formation of a photochemical smog from HC and NO_x. Emissions legislation is historically and geographically very complex, with the strictest controls in the USA and Japan; European legislation is now approaching the same levels, and this is given in table 2.2.

The emissions of CO, NO_x and HC vary between different engines, and are dependent on such variables as ignition timing, load, speed and in particular the air/fuel ratio. Typical variations of emissions with air/fuel ratio are shown in figure 2.13.

Carbon monoxide is most concentrated with fuel-rich mixtures, as there will be only partial combustion of the fuel; indeed, hydrogen will also be present in the exhaust, as shown by the 'water gas' equilibrium reaction in which carbon monoxide (CO), water vapour (H_2O), carbon dioxide (CO_2) and hydrogen (H_2) co-exist:

$$CO + H_2O \rightleftharpoons CO_2 + H_2$$

With lean mixtures, CO is still present because of dissociation. The position of the equilibrium is a strong function of temperature, and the excess air present with lean mixtures both lowers the temperature and alters the oxygen concentration in such a way as to reduce the CO emissions. In practice, the concentration of CO is greater than that predicted by equilibrium

Table 2.2 EEC legislation for passenger car emissions

Vehicle category (engine size)	Implementation dates (1 Oct.)		Pollutants (g/ECE test)		
	New models	New cars	CO	HC and NO_x combined	NO_x
2 litres and above	1988	1989	25	6.5	3.5
1.4–2.0 litres	1991	1993	30	8.0	No separate limit
Below 1.4 litres stage 1	1990	1991	45	15.0	6.0
Below 1.4 litres stage 2	1992	1993	Limits to be decided		

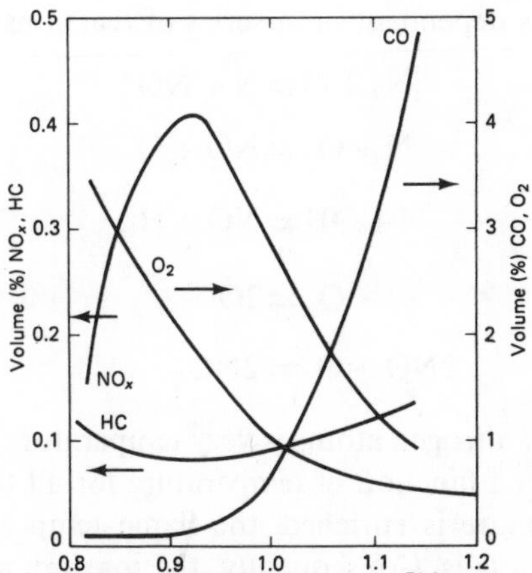

Figure 2.13 Spark ignition engine emissions for different fuel/air ratios. [Courtesy of Johnson Matthey]

thermodynamics. There is only a finite time for combustion in an engine, and this is insufficient for a thermodynamic equilibrium to be attained. The emissions of CO are largely insensitive to load and speed variations, so it is essentially the air/fuel ratio that governs the CO emission levels.

The sources of hydrocarbon emissions are reviewed by Ferguson (1986). Hydrocarbon (HC) emissions arise from unburnt fuel, and the principal sources are flame quenching in the thermal boundary layer and unburnt fuel trapped in oil films or crevices such as the piston top land and ring. Hydrocarbon emissions are reduced by excess air (fuel-lean mixtures), until the emissions rise as the flammability of the mixture reduces. Hydrocarbon emissions are reduced by any technique that extends the lean combustion limit, such as high-energy ignition, high turbulence, high swirl or a high compression ratio. Since higher engine speeds increase the turbulence, the HC emissions reduce with increasing speed; there is no significant variation in HC emisions over most of the load range.

As an engine ages, there is a build-up of deposits in the combustion chamber 'coke', and this is particularly pronounced when using fuels with lead alkyl additives. Because of fuel absorption and quenching by the combustion deposits, there is a significant increase with time in HC emissions when leaded fuel is used. Unleaded fuel shows a negligible increase in HC emissions. The HC emissions are greatest towards the end of the exhaust stroke, as the thermal boundary layer leaves the cylinder last. Variable valve timing can be used to provide a form of controlled exhaust gas recirculation, which recirculates exhaust gases with the highest concentration of HC, by controlling the extent of the valve overlap period.

Nitrogen oxides (NO_x) are a mixture of NO and NO_2, and their formation is complex, since it is dependent on a series of reactions:

$$N_2 + O \rightleftharpoons N + NO$$

$$N + O_2 \rightleftharpoons NO + O$$

$$N + OH \rightleftharpoons NO + H$$

$$O_2 \rightleftharpoons 2O$$

$$2NO + O_2 \rightleftharpoons 2NO_2$$

The oxidation rate of nitrogen atoms is very temperature dependent, and the equilibrium (which is a function of temperature for all the reactions) is not attained. As the mixture is enriched, the flame temperature rises but the oxygen concentration falls. Consequently, the maximum emissions of NO_x occur with a mixture that is slightly weak of stoichiometric.

With weak mixtures, increasing the speed reduces NO_x emissions since there is less time for their formation. As NO_x formation is highly temperature dependent, anything that reduces the combustion temperature will reduce the NO_x emissions–for example, reducing the load, retarding the ignition timing, increasing the specific humidity of the air or using exhaust gas recirculation (EGR). Between 5 and 10 per cent EGR is likely to halve NO_x emissions, but at the expense of reduced efficiency.

There are currently two main strategies for controlling engine emissions: these are lean burn engines and engines with an exhaust gas catalyst. Both these approaches are discussed in the next section on combustion system design, along with their implications on efficiency.

2.4.4 Combustion system design

There are four key considerations in combustion chamber design if the combustion is to occur efficiently.

(a) *The distance travelled by the flame front should be minimised.* By minimising the flame travel the combustion will be as rapid as possible. This has two effects: firstly it permits high engine speeds and thus high outputs, and secondly rapid combustion reduces the time in which the chain reactions that lead to self-ignition can occur. Thus, for geometrically similar engines, those with the smallest cylinders will be able to use the highest compression ratios.
(b) *The exhaust valve(s) and sparking plug(s) should be close together.* Since the exhaust valve is very hot (possibly incandescent), the flame front should propagate past the exhaust valve before it can cause pre-ignition or induce self-ignition in the unburnt gas.
(c) *There should be sufficient large-scale turbulence to promote rapid combustion.* Again, this is to reduce the time available in which self-ignition might occur. However, too much turbulence leads to excessive heat transfer from the cylinder contents, and produces combustion that is too rapid and noisy.
(d) *The gas that is burnt at the end of combustion (the 'end gas') should be in a cool part of the combustion chamber.* This is to ensure that self-ignition does not occur. There is a small clearance between the cylinder head and piston in the 'squish' area to generate turbulence; this region also produces a cool area that should coincide with the end gas.
(e) *The combustion chamber should be compact and free from crevices.* A compact combustion chamber is implicit in having a short flame travel (a). Crevices should be avoided since the flame front will be extinguished by quenching and any air/fuel mixture in a crevice will remain unburnt. This leads to a reduction in efficiency and high hydrocarbon exhaust emissions.

The surface-to-volume ratio also has a direct bearing on the compactness of the combustion chamber and, for a given geometry, the larger the swept volume the better the ratio of surface to volume. The surface-to-volume ratio is also influenced by the ratio of the stroke length to the bore diameter; a long stroke engine will produce a more compact combustion chamber. However, in the past it has been common practice to use a bore diameter greater than the stroke, since this allows larger diameter valves to be used which promote a good high-speed performance.

As with many cases in engineering there is no unique best solution, and this accounts for the numerous different combustion chamber designs that are used, all of which seem to offer a comparable performance. In practice, the choice of combustion chamber will be strongly influenced by manufacturing considerations and past experience. Since any combustion chamber design is still optimised by engine experiments, past experience is particularly important. Three combustion chamber designs that represent different approaches, and merit further individual discussion, are the Ricardo high ratio compact chamber (HRCC), the Nissan NAPS-Z and the four-valve pent-roof combustion chamber.

The three different combustion systems that are to be discussed here are shown in figure 2.14; each has been the subject of extensive tests conducted by Ricardo, and reported by Collins and Stokes (1983). The main characteristic of the four-valve pent-roof combustion chamber is the large flow area provided by the valves. Consequently there is a high volumetric efficiency, even at high speeds, and this produces an almost constant bmep from mid speed upwards.

The Nissan NAPS-Z combustion system has twin spark plugs, and an induction system that produces a comparatively high level of swirl. While the combustion initiates at the edge of the combustion chamber, the swirling flow and twin spark plugs ensure rapid combustion. With both the four-valve design and the NAPS-Z combustion chamber there is comparatively little turbulence produced by squish. In the case of the four-valve head, turbulence is generated by the complex interaction between the flows from the two inlet valves.

The high ratio compact chamber (HRCC) has a large squish area with the combustion chamber centred around the exhaust valve. The rapid combustion, which is a consequence of the small combustion chamber and high level of turbulence, allows a higher compression ratio (by 1 to 2 ratios) to be used for a given quality fuel.

A disadvantage of producing a high swirl is that the kinetic energy for the flow is obtained at the expense of a reduced volumetric efficiency. Swirl is particularly useful for ensuring rapid combustion at part load, and this leads to the concept of variable swirl control. By having twin inlet tracts, one of which is designed to produce swirl, a high swirl can be selected for part load operation. Then at full load, with the second tract open the swirl is reduced, and the volumetric efficiency is optimised. This approach can be extended to include the variable geometry induction system described in section 2.4.1.

All three engines have different combustion systems, but in each case combustion is comparatively rapid. Significant differences only appear in the combustion speed with lean mixtures, in which case the combustion speed is fastest with the HRCC, and slowest with the four-valve chamber.

The differences in specific fuel consumption need to be considered in the light of the different fuel quality requirements. Collins and Stokes determined

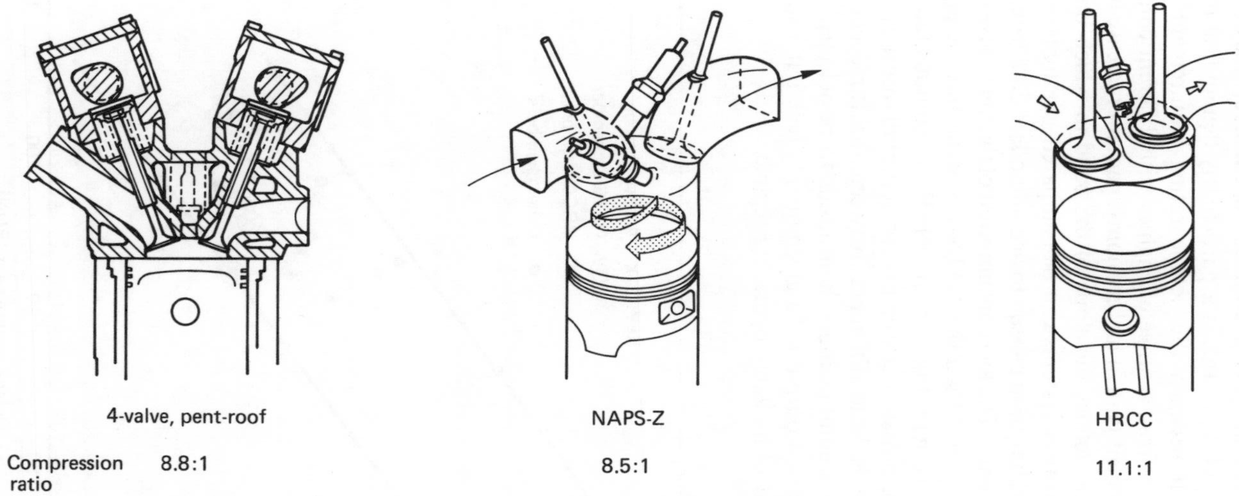

Figure 2.14 Three effective combustion systems, Collins and Stokes (1983). [Reprinted with permission © 1983 Society of Automotive Engineers, Inc.]

the optimum specific fuel consumption at 2400 rpm and part load (2.5 bar bmep); they argue that this is typical of mid-speed operation. In contrast, the octane rating requirement for each combustion chamber was determined at 1800 rpm with full load, since this is a particularly demanding operating condition. The trade-off between the octane rating requirement and the specific fuel consumption from a range of engines with different combustion systems suggests a 1 per cent gain in fuel economy, per unit increase in octane rating requirement. This defines the slope of the band in which the results from these three combustion systems can be compared–see figure 2.15.

The width of the band has been chosen to accommodate data from a range of other combustion systems. The fuel consumption of the four-valve chamber was some 6–8 per cent worse than the NAPS-Z system, but the peak level of NO_x production was half that of the NAPS-Z combustion system. However, when the NO_x emissions are compared at the full throttle maximum economy setting, there is little difference between the three combustion chamber designs. Thus a combustion system needs to be selected not only in terms of its efficiency and output for a given quality fuel, but also its level of emissions in the light of its likely operating regime.

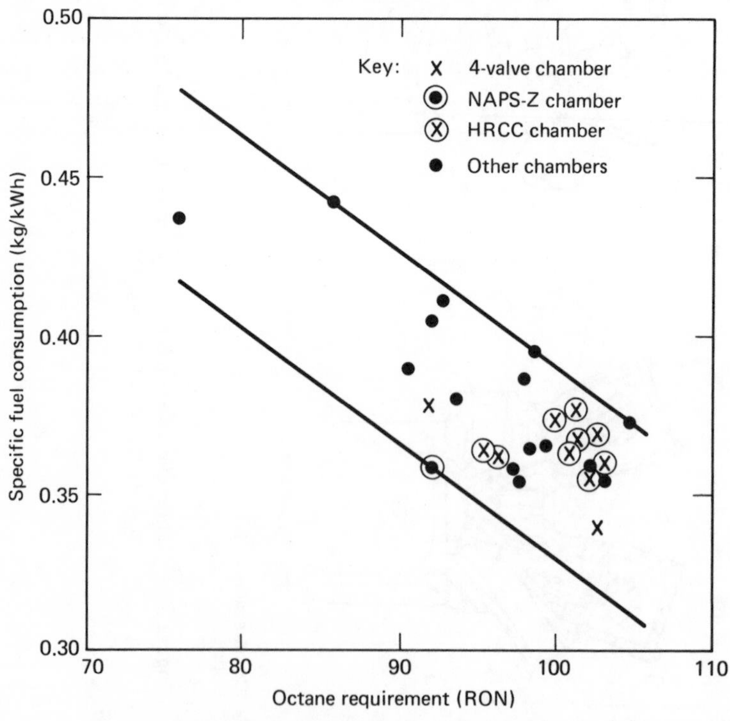

Figure 2.15 Trade-off between fuel economy and octane requirement for different combustion chambers, adapted from Collins and Stokes (1983)

In addition to the low fuel consumption of the HRCC-type system, it also allows a leaner mixture to be burnt; the equivalence ratio can be as low as 0.6, compared with 0.7 for the four-valve or NAPS-Z systems. If an engine is operated solely with a lean mixture and a high level of turbulence, then high compression ratios can be obtained with conventional fuels. The attraction of such an engine is the potential improvement in fuel economy and, more significantly, the potential for reduced emissions of CO and NO_x. However, these compact combustion chambers are prone to knock and pre-ignition, and are often limited to applications with automatic gearboxes.

Increasing the level of turbulence has several effects. Firstly, leaner mixtures can be burnt for a given level of hydrocarbon emissions, and these mixtures will be less susceptible to self-ignition because of the lower combustion temperatures. Secondly, a higher level of turbulence makes the combustion more rapid, giving less time for the chain reactions that lead to self-ignition to occur. These effects can be seen in figure 2.16, where it can also be seen that raising the compression ratio allows leaner mixtures to be burnt. This is attributable to the pressures and temperatures being higher at the end of the compression stroke, thereby making the combustion more readily self-sustaining.

The three combustion systems discussed here (the four-valve, HRCC and NAPS-Z) can all be operated with weak mixtures (lean burn) to give low emissions, thereby offering the potential to meet the European exhaust emissions legislation. The alternative approach to reducing engine emissions is to use an exhaust catalyst system. In the last decade there have been significant improvements in catalysts, culminating in the use of the three-way

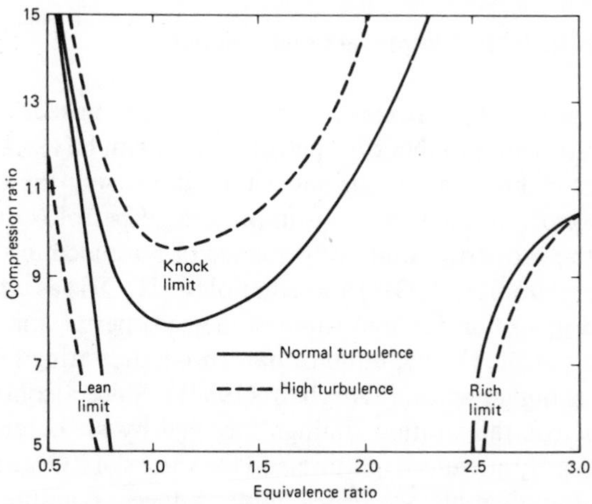

Figure 2.16 Effect of turbulence on increasing the operating envelope of spark ignition engines, adapted from Dorgham (1982)

catalyst. If a stoichiometric mixture is burnt in an engine, the three-way catalyst can oxidise the carbon monoxide (CO) and hydrocarbons (HC), and reduce the nitrogen oxides (NO_x) to produce carbon dioxide (CO_2), water vapour (H_2O) and nitrogen (N_2). However, for these reactions to occur, the air/fuel mixture needs to be kept very close to stoichiometric (± 2 per cent), and this necessitates electronic fuel injection, with a feedback signal provided by an oxygen sensor in the exhaust. This type of system is expensive, and could add 15 per cent or so to the cost of a small car; the in-service costs of catalyst replacement are yet to be fully established. In addition, an engine operating with a stoichiometric mixture will be less efficient than a lean burn engine, and this will introduce a fuel economy penalty on any vehicle; Jefferson (1985) suggests that the difference might be as much as 20 per cent.

Another complication with catalyst systems is their susceptibility to poisoning, notably from the lead compounds that originate from the lead alkyl additives that are used to raise the octane number. This was the original reason for introducing lead-free petrol, but now environmental pressures are also supporting the trend to lead-free petrol. However, there will still be a problem in Europe, since cars travel between countries, some of which may not have lead-free petrol.

Thus, a catalyst system with stoichiometric mixture control would be more expensive than a lean burn system, and have a worse fuel consumption. A catalyst system would also require lead-free petrol, and this implies either a lower compression ratio and poorer fuel economy, or a higher cost at the oil refinery. The disadvantage of a lean burn engine would be meeting the legislation requirements in the larger engine sizes.

2.5 Alcohols as fuels, fuel improvers and fuel extenders

Alcohols have properties that make them well suited to use in spark ignition engines. They are also miscible with petrol, so they can be used for extending limited reserves of fuel, and at the same time improving the octane rating. The use of alcohols as extenders and improvers of petrol is discussed here, after treating the properties and performance of pure alcohols as fuels.

The use of methanol (CH_3OH) and ethanol (C_2H_5OH) as alternative fuels was mentioned in section 1.4, and some of their properties make them more attractive than petrol. Firstly, ethanol has an octane rating (RON) of 106, and methanol is higher again [Dorgham (1982)]. Since alcohols burn more rapidly than petrol, the ignition timings specified by the octane rating tests lead to unnecessarily advanced ignition and the values of RON are misleading. Consequently the alcohols are tested under adverse conditions, and their anti-knock performance in engines is better than their octane rating would suggest. However, methanol in particular is susceptible to pre-ignition, and this is more likely to restrict the compression ratio, to limit surface temperatures in the combustion chamber.

Secondly, the alcohols have numerically lower gravimetric stoichiometric air/fuel ratios than petrol, and the alcohols have higher enthalpies of evaporation than petrol: these properties are both listed in table 2.3.

Even though smaller percentages of the alcohols evaporate during mixture preparation, the greater evaporative cooling effect is such that the alcohols produce lower mixture temperatures, with consequential improvements to the volumetric efficiency of the engine. The lower air/fuel ratios for alcohols mean that the chemical energy released per kg of stoichiometric mixture burnt during combustion is greater than for petrol, despite the lower specific enthalpies of combustion. Table 2.4 lists the enthalpy of combustion on the basis of 1 kg of a stoichiometric mixture. The improved volumetric efficiency and the higher combustion energy both increase the output of the engine, and thus reduce the significance of mechanical losses, thereby improving the overall efficiency. Goodger (1975) reports the comparisons with hydrocarbon fuels made by Ricardo at a fixed compression ratio, in which there was a 5 per cent improvement in efficiency using ethanol, and a 10 per cent improvement when methanol was used. In racing applications, methanol is particularly attractive as the power output increases with a richer mixture, up to an equivalence ratio of about 1.4 (substantially richer than for petrol).

Thirdly, the more rapid combustion of alcohols and their greater ratio of moles of products to moles of reactants, both improve the cycle efficiency. Fourthly, alcohols can operate with leaner mixtures, and this leads to lower emissions (section 2.4.4).

The disadvantages associated with alcohols are their lower volatility and energy density (tables 2.3 and 2.4), their miscibility with water, and a tendency towards pre-ignition. The low volatility can lead to difficult cold starting (especially below 10°C), and heated inlet manifolds are required for normal operation.

Table 2.3 Properties of alcohols that affect mixture preparation

Fuel	Boiling point (°C)	Stoichiometric gravimetric air/fuel ratio	Enthalpy of evaporation (kJ/kg)
Methanol	65	6.5	1770
Ethanol	78.5	9.0	850
Petrol	25–175	14.5	310

Table 2.4 Enthalpies of combustion for alcohols and petrol

Fuel	Enthalpy of combustion	
	(MJ/kg fuel)	(MJ/kg stoichiometric mixture)
Methanol	22.2	3.03
Ethanol	29.7	2.97
Petrol	42.0	2.71

Alcohols are attractive as additives (or more correctly extenders) to petrol, since the higher octane rating of the alcohols raises the octane rating of the fuel. Goodger (1975) reports that up to 25 per cent ethanol yields a linear increase in the octane rating (RON) of 8 with 92 octane fuel and an increase of 4 with 97 octane fuel.

Dorgham (1982) reports on extensive trials conducted by Ford that encompassed pure alcohols, and mixtures with petrol. In Brazil, where there is an ethanol fuel policy, Ford has increased the compression ratio of production cars to 13:1. In Europe, blends of petrol with 3 per cent alcohol have been commonly used to improve the octane rating, and West Germany has sponsored a programme to introduce 15 per cent of methanol into petrol.

The quantities of alcohol and other oxygenated fuels that the EEC allows to be added to petrol are summarised by Palmer (1986). Palmer also reports on vehicle tests, which showed that all oxygenate blends gave a better anti-knock performance during low speed acceleration than hydrocarbon fuels of the same octane rating. Furthermore, there is a tendency for the anti-knock benefits of oxygenate fuels to improve in unleaded fuels. However, care is needed with ethanol blends to avoid possible problems of high-speed knock. Fuel consumption on a volumetric basis is almost constant as the percentage oxygenate fuel in the blend increases. The lower calorific value of the oxygenates is in part compensated for by the increase in fuel density. None the less, Palmer (1986) shows that energy consumption per kilometre falls as the oxygenate content in the fuel increases.

Ethanol is entirely miscible with petrol, while methanol is only partially miscible. The miscibility of both alcohols in petrol reduces with the presence of water and lower temperatures. To avoid phase separation, as moisture becomes absorbed in the fuel, chemicals such as benzene, acetone or the higher alcohols can be added to improve the miscibility. Palmer (1986) also reports on materials compatibility, hot and cold weather drivability, altitude effects, and the exhaust emissions from oxygenate blends.

2.6 Concluding remarks

To produce a fuel-efficient engine in which the part load fuel consumption is also low, it is essential to have an engine with a high mechanical efficiency. This is achieved by careful attention to detail, such as balancing and the sizes of bearings. It is equally important to ensure that all the engine-driven ancillary items are efficient, because any deficiencies here will also have a detrimental effect on the fuel economy, especially at part load.

The compression ratio also has a significant influence on the fuel economy at all operating points. If there was no limitation on fuel quality, the optimum compression ratio might be in the range 14–17:1; the exact value will depend on the mechanical efficiency of the engine. In practice, of course, the fuel quality is limited, and this will influence both the type of combustion system and the compression ratio. The final selection of a combustion chamber will also be influenced by previous experience, emissions legislation, cost and the ease of manufacture.

High efficiency spark ignition engines are likely to have an open combustion chamber, of the type described in section 2.4.4. The engines will have a high compression ratio, for the combustion of lean air/fuel mixtures. The induction system is likely to have variable geometry features that provide induction tuning and the control of swirl. Variable valve timing is also likely to be used to reduce the fuel consumption and emissions (see section 2.4.1).

The best economy is likely to be shown by engines with electronic engine management systems, for the control of the fuelling and ignition (section 2.4.2). When low emissions are also important, engines are likely to have multi-point injection, with control of the phasing between the start of injection and inlet valve opening. The ignition timing will be optimised most successfully on engines that retard the ignition when knock is sensed, whether the knock is caused by low-quality fuels or engine ageing. Ignition is likely to be by high-energy sparks, which are produced by separate high-voltage systems for each cylinder.

Large-capacity engines (above about 2.0 litres) are likely to need exhaust gas catalysts to reduce the emissions to acceptable levels (table 2.1). Such systems increase the engine cost, and are less efficient as a consequence of burning stoichiometric mixtures.

While the four stroke reciprocating spark ignition engine continues to dominate automotive applications, work is still continuing on the Wankel and two stroke engines. The Wankel engine suffers from a poor combustion chamber geometry, mechanical sealing, local overheating and friction problems. The attractions of the two stroke engine are its higher specific output and mechanical simplicity. Unfortunately, poor control over the gas exchange process has traditionally led to a poor fuel economy and high emissions. However, Barrie (1986) reports that the Orbital Engine Company of Australia now claims to have overcome these problems. The three-cylinder

Orbital two stroke engine has over 200 fewer parts than a comparably powerful 1.6-litre four-cylinder four stroke engine, and a 68 per cent weight saving. By using an equivalence ratio of less than 0.5, it is claimed that fuel consumption can be reduced by as much as 20 per cent. However, an oxidation catalyst is required in the exhaust system to remove the unburnt hydrocarbons. The weight saving in itself is significant, since this could amount to a 10 per cent reduction in the vehicle weight, and as discussed in chapter 6, this could lead to a fuel economy improvement of a few per cent.

However, it seems unlikely that the four stroke reciprocating engine will be displaced quickly, if for no other reason than the high investment that is associated with existing engine designs.

2.7 Example

An engine with a rated output of 45 kW is cooled by a pressed steel fan that has an efficiency of 12.5 per cent and absorbs 2.1 kW at the rated engine speed. Calculate the improvement in engine performance if the fan is replaced by a moulded plastic fan with an efficiency of 33 per cent; state clearly any assumptions that are made.

Solution

Assume that the output from the engine is a net value, and that the fan drive efficiency is included as a component of the fan efficiency.

Find the reduced power consumption of the new fan, $\dot{W}_f$:

$$\dot{W}_f = 2.1 \times \frac{0.125}{0.33} = 0.8 \text{ kW}$$

The net output ($\dot{W}_b$) at the rated speed and load is now

$$\dot{W}_b = 45 + (2.1 - 0.8) = 46.3 \text{ kW}$$

There is thus a 1.3 kW or 2.9 per cent improvement in the net power output at the rated speed and load. At this condition the fuel supply rate to the engine can be assumed to be the same. Thus there is a 2.9 per cent improvement in power output for the same fuel consumption; in other words there is a 2.9 per cent improvement in fuel economy.

Improvements in the fuel consumption will be most apparent at part load. Assume, here, that at part load operation the fuel supply rate of the engine is reduced to maintain the same power output, and further assume that the efficiency of the engine is constant for these small changes in fuelling. The results are summarised in table 2.5. If the fan efficiency remains constant, the power absorbed will be proportional to the speed cubed. Thus at half the

Table 2.5 Improvement in part load fuel consumption at the rated speed

Output $\dot{W}_b$ (kW)	Output at the same fuel supply rate with the new fan $\dot{W}_n$ (kW)	Improvement in fuel economy $\frac{\dot{W}_n - \dot{W}}{\dot{W}_n} \times 100$ (%)
22.5	23.8	5.5
15.0	16.3	8.0
4.5	5.8	22.4

rated speed the power consumption and saving will only be 1/8 of the values at the rated speed.

The arguments applied here are equally applicable to both compression ignition and spark ignition engines. Furthermore, a reduction in the power absorbed in the engine by friction can be treated in exactly the same way.

2.8 Discussion points

(1) How does the compression ratio and the air/fuel ratio affect engine efficiency and output?
(2) What is the octane rating a measure of, and how can it be improved in the absence of the lead alkyl additives?
(3) How can the part load efficiency of spark ignition engines be improved?
(4) What are the requirements of the gas exchange processes, and how can they be achieved?
(5) Itemise the requirements of the ignition system.
(6) What are the advantages of a multi-point timed sequential fuel injection system?
(7) Why are alcohols attractive as fuels for spark ignition engines?
(8) What are the requirements for an efficient combustion chamber?
(9) What are the effects of the air/fuel ratio, load, speed and ignition timing on the emissions of unburnt hydrocarbons, carbon monoxide and nitrogen oxides?
(10) Explain how the emissions of nitrogen oxides can be reduced.

3 Diesel Engine Fuel Economy

3.1 Introduction

As with the previous chapter on spark ignition engines, the purpose of this chapter is to highlight the aspects of diesel engine design that influence efficiency.

In any comparison of diesel with spark ignition engines, the first thing to remember is that the energy content of the fuels is different, on both a volumetric and gravimetric basis. If a diesel engine and spark ignition engine had the same (gravimetric) specific fuel consumption, then the diesel engine would be the more efficient, as diesel fuel has the lower calorific value. Conversely, if a spark ignition and diesel engined vehicle had the same fuel consumption (litres/100 km), then it would be the petrol engined vehicle that was more efficient, since petrol has the lower volumetric calorific value. Furthermore, the energy cost at the refinery is generally greater for producing petrol than diesel fuel. Typical energy contents and refinery costs are shown in table 3.1.

At full load, diesel engines are more efficient than spark ignition engines, and at part load, the efficiency of diesel engines falls less rapidly than for petrol engines; the reasons are discussed in the next section. None the less, the frictional losses and power consumed by the ancillaries are just as important in diesel engines as in petrol engines, and the material developed in section 2.3 is entirely relevant. Unfortunately, the air utilisation in diesel

Table 3.1 Comparison of the energy contents of petrol and diesel, from Francis and Woollacott (1981)

Fuel	Gravimetric calorific value (MJ/kg)	Density (kg/m^3)	Volumetric calorific value (MJ/litre)	Refinery cost (MJ/kg)	Primary energy cost (MJ/kg)
Petrol	44	795	31.8	2.7	46.7
Diesel	42	910	38.15	1.65	43.65

engines is less effective than in spark ignition engines, and this results in a lower torque output for a given size of engine.

Another limitation of diesel engines is that their speed range is more restricted than that of spark ignition engines. This imposes a further restriction on the output of diesel engines; for naturally aspirated engines, an output 20 kW/litre is quite typical while a spark ignition can easily achieve 40 kW/litre. The speed range of small diesel engines is extended by using indirect injection, but this impairs the efficiency for the reasons discussed in section 3.3.

Fortunately, turbocharging is a very attractive means of increasing the output of diesel engines. An additional advantage is that since the mechanical losses do not rise in direct proportion with the engine output, the overall efficiency will also be improved. As combustion in a diesel engine is initiated by self-ignition, turbocharging improves combustion–the exact opposite to what happens when turbocharging petrol engines.

Another development that offers scope for improving the efficiency of diesel engines is the low heat loss engine or so called 'adiabatic' engine. In a spark ignition engine, any attempt to reduce the level of cooling is likely to lead to a combination of self-ignition and pre-ignition, which would usually lead to a reduction in output and the failure of components. In contrast, with a diesel engine, the reduced heat transfer would aid ignition and combustion. The reduced heat transfer would lead to higher pressures and temperatures during the expansion stroke, thereby increasing the expansion work. More significant gains would be found in a turbocharged engine, since the turbine entry temperature and pressure would be higher. In both cases the reduced load on the coolant system leads to useful savings in the cooling fan and pump. This is particularly helpful in maintaining a good part load efficiency, as is any means of reducing frictional losses (as was discussed in section 2.3).

The electronic control of fuel injection, in particular the timing of injection, will enable engines to be optimised for high output, efficiency, low noise and low emissions, under conditions of varying load, speed, engine temperature and turbocharger boost pressure. Fortunately, the emissions from diesel engines are inherently low, because the air/fuel ratio is always weak of stoichiometric. However, exhaust smoke is the usual limitation on diesel engine output; when the fuelling rate increases there is insufficient time for complete mixing and combustion of the fuel.

3.2 Essential thermodynamics

3.2.1 Diesel cycle analysis

The simplest model for the diesel engine is the diesel cycle, which shows how the efficiency varies with the compression ratio and the load ratio:

$$\eta_{\text{Diesel}} = 1 - \frac{1}{r_v^{\gamma-1}}\left[\frac{\alpha^\gamma - 1}{\gamma(\alpha - 1)}\right] \tag{3.1}$$

where r_v = volumetric compression ratio

γ = ratio of the specific heat capacities for air, c_p/c_v

and α = load ratio, the volumetric expansion ratio that is assumed to occur at constant pressure during the heat addition.

Comparison with the Otto cycle efficiency (equation 2.1) shows that the only difference is the term in square brackets; as before, the efficiency should rise as the compression ratio is increased. In this simple model, combustion is assumed to start at top dead centre and continue at constant pressure to αV_c (V_c is the clearance volume at top dead centre). The term in square brackets is always greater than unity, and the limiting case, as the output is reduced, is that $\alpha = 1$, and the term in square brackets also reduces towards unity. The effect of this on cycle efficiency is shown in figure 3.1.

The load ratio normally has a value greater than unity, and this might suggest that the efficiency of a diesel engine might be lower than a spark ignition engine. However, there are three reasons why the cycle efficiency of the diesel will be higher. Firstly, the compression ratio of a spark ignition engine is usually less than 10:1, while the compression ratio of a diesel engine is normally in the range 12–24:1.

Secondly, since the diesel engine is initially just compressing air, and is subsequently burning a weak mixture, the efficiency will be greater than that shown for the fuel/air cycle efficiency of a spark ignition engine for the same

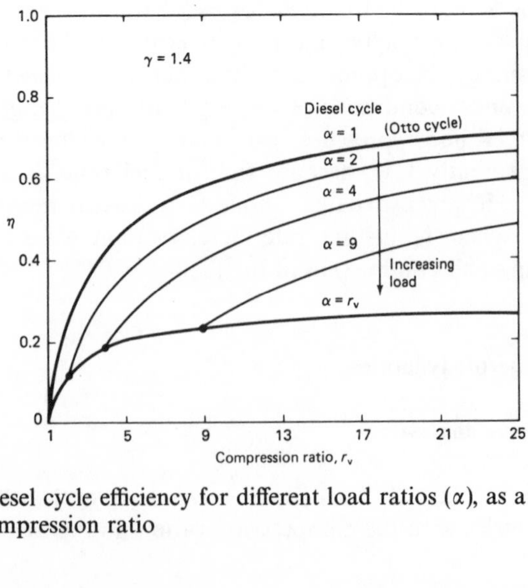

Figure 3.1 Diesel cycle efficiency for different load ratios (α), as a function of the compression ratio

equivalence ratio (figure 2.1). Thirdly, in any comparison of spark ignition and diesel engines with the ideal Otto and Diesel cycles, it must be remembered that combustion in a spark ignition engine does not occur solely at constant volume, nor does combustion occur solely at constant pressure in a diesel engine. A better representation of the combustion process is given by the Dual cycle. In the Dual cycle the heat addition is divided between constant-volume and constant-pressure processes, and at a given compression ratio the efficiency lies between that calculated by the Otto and Diesel cycles. To summarise, the diesel engine has a higher maximum efficiency than a spark ignition engine for three reasons:

(a) The compression ratio is higher.
(b) During the initial part of compression, only air is present.
(c) The air/fuel mixture is always weak of stoichiometric.

Furthermore, the diesel engine is, in general, designed to operate at lower speeds, and consequently the frictional losses are smaller.

In a diesel engine the air/fuel ratio is always weak of stoichiometric, in order to achieve complete combustion. This is a consequence of the very limited time in which the mixture can be prepared. The fuel is injected into the combustion chamber towards the end of the compression stroke, and

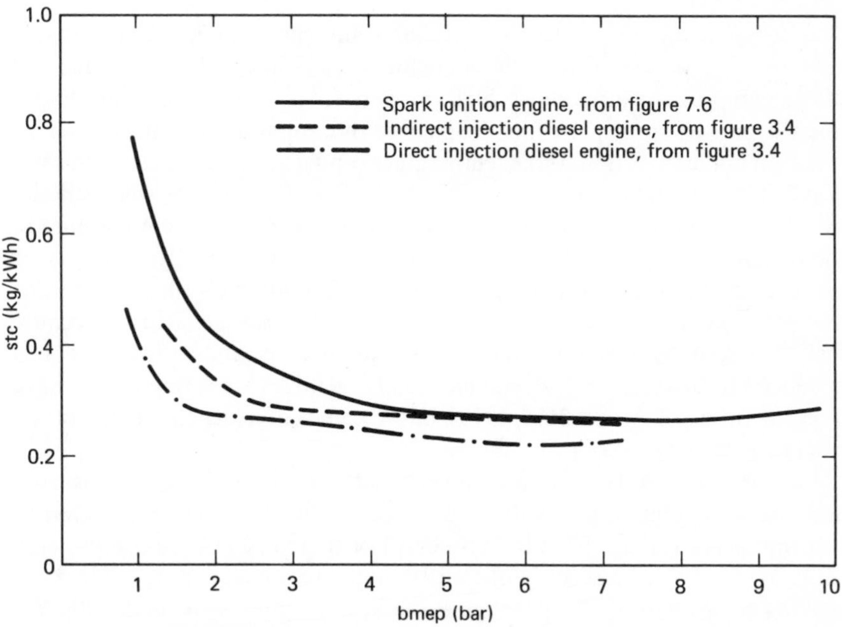

Figure 3.2 Comparison of the part load efficiency of spark ignition and diesel engines at 2000 rpm

around each droplet the vapour will mix with air, to form a flammable mixture. Thus the power can be regulated by varying the quantity of fuel injected, with no need to throttle the air supply.

As shown in figure 3.2, the part load specific fuel consumption of a diesel engine rises less rapidly than for a spark ignition engine. A fundamental difference between spark ignition and diesel engines is the manner in which the load is regulated. A spark ignition engine always requires an air/fuel mixture that is close to stoichiometric. Consequently, power regulation is obtained by reducing the air flow as well as the fuel flow. However, throttling of the air increases the pumping work that is dissipated during the gas exchange processes.

Also, since the output of a diesel engine is regulated by reducing the amount of fuel injected, the air/fuel ratio weakens and the efficiency will improve. Finally, as the load is reduced, the combustion duration decreases, and the cycle efficiency improves. To summarise, the fall in part load efficiency is moderated by:

(a) The absence of throttling.
(b) The weaker air/fuel mixtures.
(c) The shorter duration combustion.

3.2.2 Diesel engine combustion

The compression ratio also has a significant effect on ignition, and the subsequent combustion in a diesel engine. Fuel is injected towards the end of the compression stroke at a high pressure (300–1000 bar), to produce a finely atomised spray. The fuel droplets evaporate and mix with the air by diffusion and convection. If the temperature (and, to a lesser extent, pressure) is sufficiently high for long enough, the fuel will self-ignite. The susceptibility of a fuel to self-ignition is indicated by its cetane rating–a counterpart to the octane rating of spark ignition engine fuels. Unfortunately, as a greater percentage of crude oil is converted to lighter fraction fuels (diesel, kerosene and petrol) there is a fall in the cetane rating of diesel fuel. This is because fuel produced by catalytic cracking has a lower cetane rating than fuel produced by distillation. Automotive diesel fuel in Britain is required to have a cetane rating of 50 (BS 2689:1970), but in some countries the cetane rating has fallen to below 40.

Fuel is least likely to ignite during starting, since the engine is cold and the slow cranking speeds lead to convective cooling of the charge, and significant leakage from the cylinder. For these reasons the compression ratio is likely to be set sufficiently high to ensure a reliable cold starting performance. The problem is worst in engines with small cylinders (< 500 cm^3), and the compression ratio is often set higher than is optimal for either maximum output or efficiency, to ensure reliable starting. In contrast, highly turbocharged engines have a peak firing pressure limitation of about 150 bar, and the compression ratio has to be reduced to a level that can make starting and light load operation difficult. These problems and some of the possible solutions, such as variable compression ratio pistons, are discussed by Ladommatos and Stone (1986).

Ignition delay, the period between the start of injection and the start of combustion, has a direct effect on combustion noise. During the ignition delay, fuel is being injected but not burnt. None the less, as the air and fuel are being heated by compression, a flammable mixture is being formed; and when combustion does occur it will do so throughout this flammable mixture. The ensuing rapid combustion leads to a rapid pressure rise, and the characteristic diesel knock. The next phase, diffusion combustion, is controlled by the rate of diffusion of the reactants into the combustion zone, and the diffusion of the products away from the reaction zone. The final stage of combustion occurs during expansion, and is the oxidation of any partial combustion products that were produced earlier.

3.3 Direct and indirect injection diesel engines

The two combustion systems shown in figure 3.3, for direct injection (DI) and indirect injection (IDI) engines, are both commonly used in diesel engines. In both types of engine, satisfactory performance depends on complete, controlled and rapid combustion.

The direct injection engine obtains adequate mixing of the fuel and air, by using a high-pressure injection system and an ordered air motion. The fuel is injected at pressures as high as 1000 bar, through injectors with multiple hole nozzles. The matching of the air and fuel motion is critical for satisfactory operation of direct injection engines. The most important form of air motion is swirl: the ordered rotation of air about the cylinder axis. Swirl is generated in the induction process by devices such as shrouded inlet valves and helical inlet ports. Unfortunately, swirl is produced by a pressure difference that can only be generated by a fall in the volumetric efficiency. Increasing the swirl in an engine with a constant fuelling rate:

(a) Makes the combustion more rapid and complete (improving efficiency and output).
(b) Increases the heat transfer (reducing efficiency and output).
(c) Reduces the volumetric efficiency (reducing efficiency and output).
(d) Increases combustion noise, as a result of the more rapid combustion.

Clearly, since swirl has both benefits and disadvantages, there is an optimum swirl ratio for any engine. In engines with low levels of swirl the adequate mixing of fuel and air is obtained by using multi-hole injector nozzles.

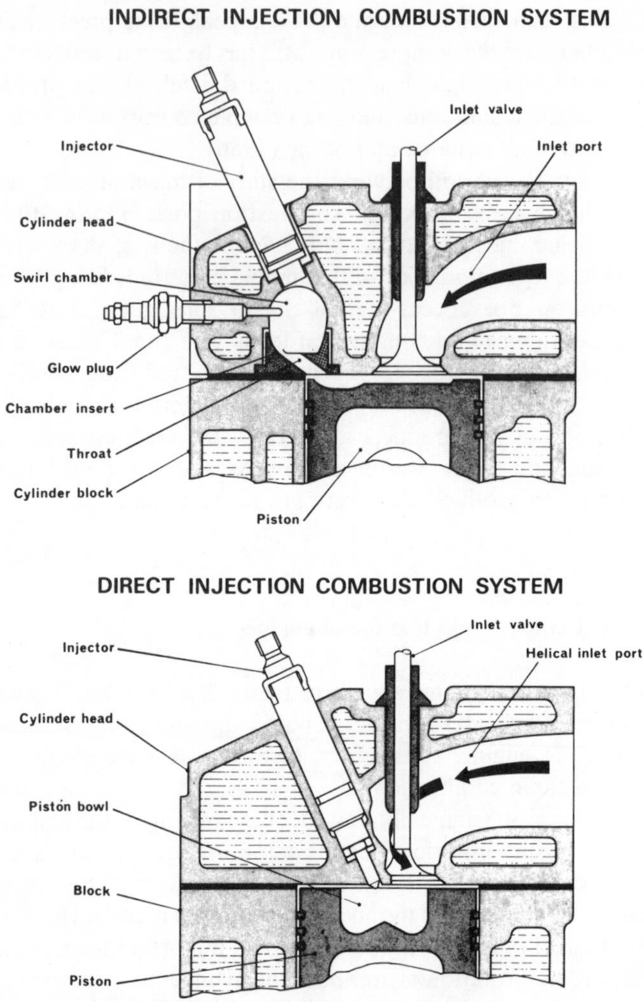

Figure 3.3 Comparison of direct and indirect injection combustion systems. [Reproduced by permission of Ford Motor Co. Ltd]

Unfortunately, multi-hole injectors require higher fuel supply pressures, for a given rate of injection and spray penetration.

While many different combustion chamber geometries exist, the common features are a bowl in the piston crown and a small clearance between the piston and the cylinder head. As the piston approaches the end of the compression stroke, the small clearance with the cylinder head produces squish, an inward turbulent flow of air. While turbulence cannot enhance the initial stages of combustion, it will help to ensure that the final stages are rapid and complete. The bowl provides a compact combustion chamber, and because of the conservation of the moment of momentum, the swirl speed in the bowl is greater than that in the cylinder during induction and compression.

Until recently, it has not been possible to devise combustion systems for direct injection engines that enable the engine to run above about 3000 rpm. This is not a limitation in large engines, since there is also a limit of about 12 m/s on the maximum mean piston speed. At 3000 rpm, the limitation of 12 m/s on mean piston speed limits the engine stroke (L) to

$$L = \frac{12}{2 \times 3000/60} = 0.12 \text{ m}$$

In other words, if the piston stroke is greater than 0.12 m, the limitation on engine speed will be the piston speed, not the combustion speed of a direct injection combustion system. Direct injection engines usually have a stroke that is greater than the bore diameter (under-square), in order to produce a compact combustion chamber. A stroke of 0.12 m would thus typically mean a swept volume of 1 litre/cylinder.

In smaller-capacity engines, restricting the maximum speed to 3000 rpm would place an unacceptable limit on engine outputs for automotive applications. The indirect injection (IDI) combustion system allows engine speeds of 5000 rpm or more, and better air utilisation than the DI system. Unfortunately, there is a fuel consumption penalty of 10–15 per cent, for reasons that will be shown shortly.

The particular form of indirect injection system shown in figure 3.3 is a swirl chamber; the best known example of this type is the Ricardo Comet combustion chamber that dates back to the 1930s. In all indirect injection engines, the fuel is injected into a pre-chamber, which is connected by a restricted passage (or throat) to the main chamber. The swirl passage is characterised by the throat being tangential to the pre-chamber. During compression, the throat generates a high-velocity flow in the pre-chamber, which ensures rapid combustion of the fuel and good air utilisation. The chamber insert or hot plug is made from a heat-resistant material of low thermal conductivity. This ensures that once the engine is operating, ignition delay is minimised: the insert temperature can rise as high as 700°C. Since the swirl speed is very high, and some of the fuel evaporates after impinging on the hot insert, the fuel injection requirements are less demanding than for a direct injection engine. Injection pressures are typically 350 bar through a single hole nozzle: the use of lower injection pressures also allows the use of the cheaper rotary injection pumps.

The higher speed capability of the IDI diesel (when compared with the DI), is due to:

(a) Heat transfer from the pre-chamber throat reducing the ignition delay.

(b) The intense air motion in the pre-chamber providing rapid mixing of the fuel and air to promote fast combustion.
(c) The comparatively low injection pressures associated with a single hole nozzle, even with the high injection rates associated with the maximum power operating point.

The disadvantages of the indirect injection engine are centred around the high heat transfer coefficients in the throat and pre-chamber–a consequence of the high flow velocities. In the first instance, starting is very difficult; compression ratios as high as 24:1 and glow plugs are used, to provide compression temperatures and pressures that are adequate for ignition. Such high compression ratios are higher than is optimal for either maximum efficiency or output, since the incremental gains in indicated efficiency are more than offset by the increased mechanical losses stemming from the higher pressure loadings. Secondly, the high heat transfer in the IDI engine during normal operation reduces the indicated efficiency and increases the load on the cooling system.

Since DI engines are more efficient than IDI engines, recent work has resulted in the speed range of direct injection engines being extended to over 4000 rpm. The first mass-produced engine of this type was the Ford 2.5 DI engine, described by Bird (1985). The high-speed operation was obtained by meticulous attention to the design of the induction system, and selective assembly of components during manufacture. All the assembled cylinder heads and inlet manifolds are checked for the correct swirl speed after assembly. Great care is also taken to avoid unacceptable stack-ups of tolerances in all the components associated with the combustion system. This engine is discussed further as part of the case study in section 7.3.2.

A direct comparison between 2.5 litre IDI and DI engines is made by Hahn (1986), and the specific fuel consumption results are shown in figure 3.4. The improvement in fuel consumption offered by the DI engine is better than 10 per cent at nearly all loads and speeds, and as high as 20 per cent with high-speed light load operation.

A very detailed description of the development of a DI engine is given by Takeuchi *et al.* (1985). One of the many variables discussed was the effect of varying the compression ratio. Figure 3.5 shows that raising the compression ratio from 18 to 19:1 led to a decrease in hydrocarbon (HC) emissions, but to an increase in the noise level. The analysis of pressure diagrams showed that the higher compression ratio led to more rapid combustion and higher peak pressures. Furthermore, the higher compression ratio produced a worse fuel efficiency and higher exhaust smoke emissions, these are listed in table 3.2. Emissions are discussed more fully in section 3.5.1.

While the output of all these 2.5 litre DI engines is directly comparable with similar displacement indirect injection engines (at about 20 kW/litre), the comparison with spark ignition engine outputs is still very poor. A

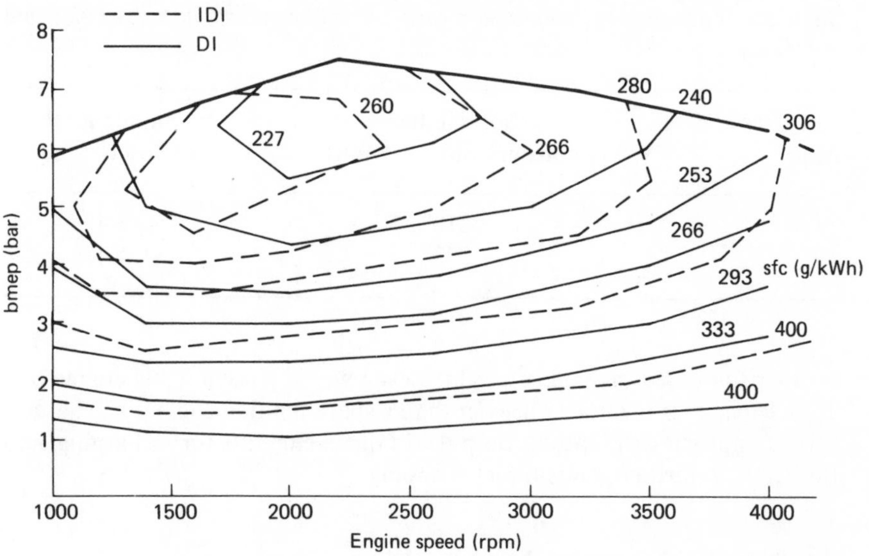

Figure 3.4 Comparison of specific fuel consumption (g/kWh) maps for four-cylinder 2.5-litre naturally aspirated DI and IDI engines, adapted from Hahn (1986)

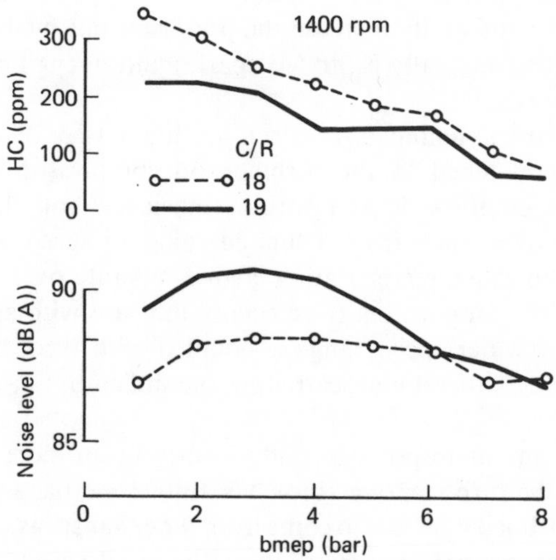

Figure 3.5 Effect of compression ratio on noise and hydrocarbon emissions, Takeuchi *et al.* (1985). [Reprinted with permission © 1985 Society of Automotive Engineers, Inc.]

Table 3.2 The effect of compression ratio on smoke and sfc at 8 bar bmep and 2000 rpm

Compression ratio	Specific fuel consumption (g/kWh)	Smoke level (Bosch No.)
18	218	3.4
19	232	5.4

comparison based on engine weight is even worse, since a diesel engine of a given capacity is invariably heavier than a spark ignition engine. The natural way of improving the specific output of a diesel engine is turbocharging, and this leads to further gains in fuel economy.

3.4 Turbocharging and low heat loss engines

3.4.1 Turbocharging

With turbocharging, air is compressed in a compressor driven by an exhaust gas turbine; the purpose is to increase the power output of the engine. By having a larger flow rate of air, more fuel can be burnt–and this leads directly to an increase in the power output. In automotive applications, the use of turbocharging can quite easily increase the power output by 40 per cent, to perhaps 28 kW/litre (naturally aspirated spark ignition engines can readily achieve 40 kW/litre).

Turbochargers are dynamic flow effect machines that rely on the gas flow angles being matched to the turbine and compressor blade angles. Consequently, the compressor and turbine efficiency both fall, once the operating point moves away from a limited range of speed and flow rate combinations. In contrast, reciprocating engines operate over a wide speed range (and thus flow range); this is especially the case with spark ignition engines, since their wider speed range is coupled with throttling to widen further the flow range, thereby exacerbating the difficulties in turbocharger matching.

Turbochargers cannot respond instantaneously to an increase in engine load or speed, as the turbocharger rotor has a finite inertia, and the energy to accelerate the turbocharger has to come from the exhaust gas. Furthermore, if the turbocharger is sized for operation at the rated speed and load, the boost pressure at low engine speeds would be inadequate. Consequently, turbochargers are selected that will give the necessary boost at low engine speeds, and then, to prevent overspeeding of the rotor at high engine speeds, the exhaust is allowed to bypass the turbine through a waste gate. For all these reasons, the selection and matching of turbochargers to reciprocating engines is difficult. A fuller discussion can be found in Stone (1985), while Watson and Janota (1982) provide a very rigorous and comprehensive treatment.

With spark ignition engines, turbocharging is usually associated with obtaining an increased output at the expense of fuel economy. With a given production facility, or space in an engine compartment, turbocharging is an expedient means of producing a high performance vehicle. As previously mentioned, the very wide flow range and rapid response needed in spark ignition engines make turbocharger matching particularly difficult. Furthermore, the increased pressure and temperature of the incoming fuel and air mixture will lead to self-ignition (section 2.2.1) unless design changes are made. The usual solutions with spark ignition engines are a combination of reduced compression ratio, retarded ignition timing and the use of a rich mixture at full load, all of which lead to a reduction in efficiency.

In contrast, turbocharging leads to an improvement in the efficiency of diesel engines at full load, since the frictional losses do not rise in direct proportion to the output. Furthermore, it is not necessary to reduce the compression ratio as in a spark ignition engine to reduce the risk of self-ignition. Indeed, self-ignition is fundamental to combustion in compression ignition engines.

Figure 3.6 shows a comparison of efficiency between a turbocharged and naturally aspirated direct injection engine (the same engine is compared with an IDI engine in figure 3.4). The torque characteristic is now determined principally by the way the turbocharger has been matched. By arranging for the maximum torque to occur at a low engine speed, good torque back-up (the negative gradient of the torque curve) is obtained. This leads to good driveability, since in any fixed gear the available torque increases as the vehicle slows down. The torque characteristics can be further improved by the use of variable geometry turbochargers. Many manufacturers are currently involved with the development of these, and an early example is provided by Isuzu (1986).

A typical automotive turbocharger is shown in figure 3.7. While radial flow compressors and turbines have lower maximum efficiencies than their axial flow counterparts, for the comparatively low flow rates in automotive applications, the radial flow machines are more efficient. The isentropic efficiency of the radial flow compressor will be mostly in the range of 60–75 per cent. A low efficiency not only increases the work required to drive the compressor but also increases the temperature rise in the compressor, thereby reducing the air density at exit. Since the air mass flow rate is dependent on the breathing capability of the engine and the density of the air in the inlet manifold, then the density ratio across the compressor is more significant than the pressure ratio. The relationship between pressure ratio

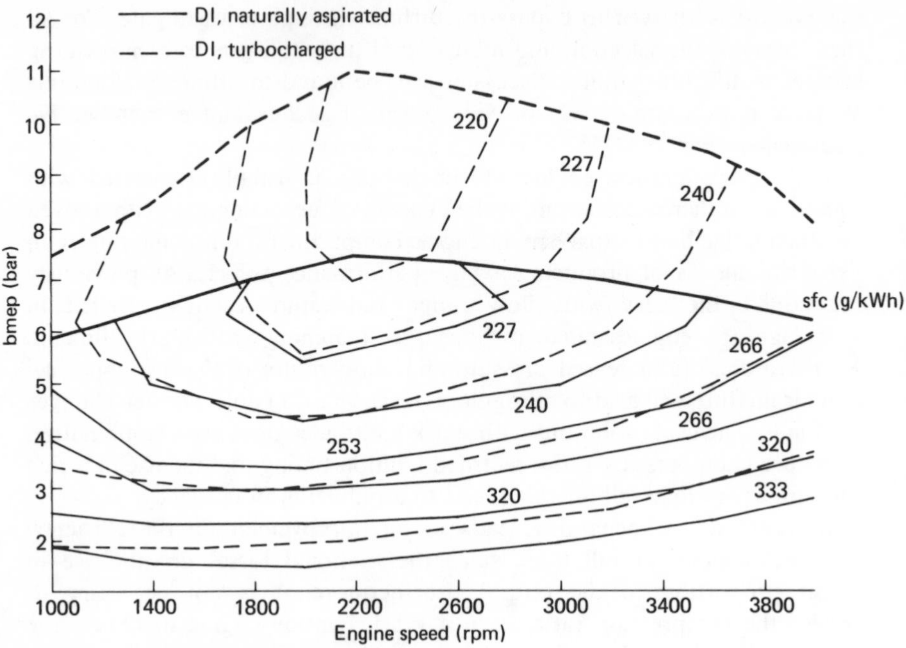

Figure 3.6 Comparison of specific fuel consumption maps (g/kWh) for a four-cylinder naturally aspirated and turbocharged 2.5 litre DI engine, adapted from Hahn (1986)

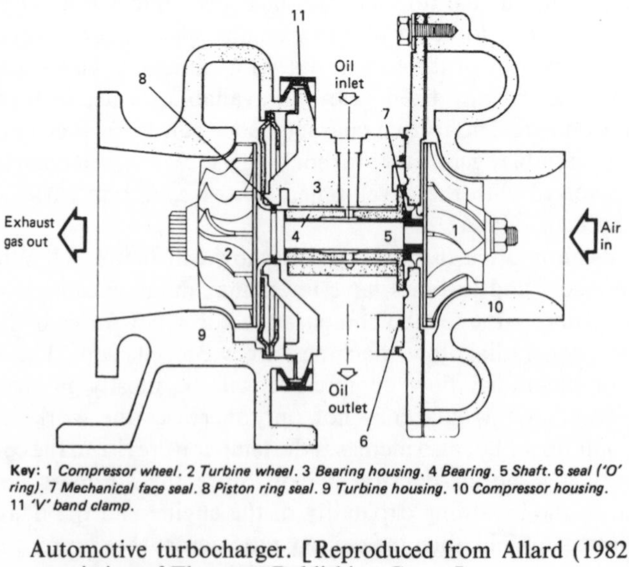

Figure 3.7 Automotive turbocharger. [Reproduced from Allard (1982) by kind permission of Thorsons Publishing Group]

and density ratio is shown in figure 3.8 for a range of compressor efficiencies. Also shown is the effect of an inter-cooler between the compressor and the engine, which can cool the air back to ambient temperature, and an inter-cooler of more realistic effectiveness (0.6). For a typical automotive pressure ratio of 2.0, there can be seen to be some advantage associated with an inter-cooler. However, if the inter-cooler is cooled by engine coolant, the air can never be cooled to ambient temperature; while if the inter-cooler is air cooled, there will be very bulky ducting.

When inter-cooling is used, not only is the temperature reduced during induction, but it is also reduced throughout all the subsequent processes. This leads to reduced heat transfer and lower thermal loadings on the piston and exhaust valve, and on the exhaust turbine. The comparatively low boost pressure in automotive engines means that the compression ratio is not reduced much (if indeed at all), to limit peak pressures. In medium and slow speed diesel engines, the larger-diameter cylinders and higher turbocharger pressure ratios (3 or even 4:1) mean that the compression ratio is reduced. The compression ratio can be as low as 12:1, to limit the peak pressures to

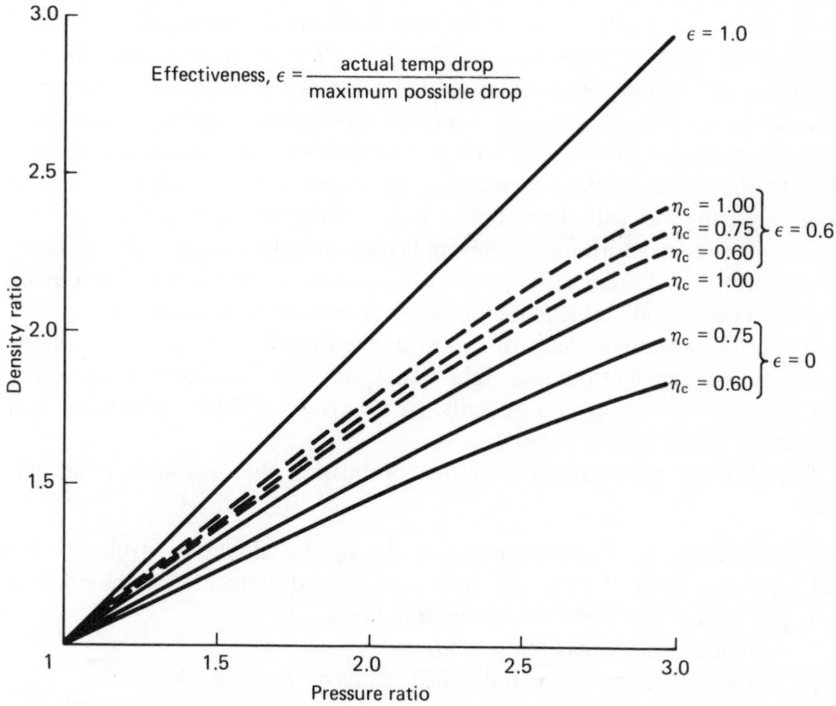

Figure 3.8 The effect of compressor efficiency and intercooler effectiveness on density ratio as a function of pressure ratio

about 150 bar. These developments and their associated problems are reviewed by Ladommatos and Stone (1986), along with some solutions.

3.4.2 Low heat loss engines

So-called adiabatic engines, or more accurately low heat loss engines, offer two types of improvement to diesel engine efficiency. Firstly, and most widely quoted, is the improvement in expansion work and the higher exhaust temperature. This leads to further gains when an engine is turbocharged. Secondly, the reduced cooling requirements allow a smaller capacity cooling system. The associated reduction in power consumed by the cooling system is, of course, most significant at part load (section 2.3.2).

Reductions in heat transfer from the combustion chamber can be obtained by redesign with existing materials, but the greatest potential here is offered by ceramics. Ceramics can be used as an insulating layer on metallic components, or more radically as a material for the complete component.

Differences between the thermal coefficients of expansion of metals and ceramics mean that great care is needed in the choice of the ceramic and the metal substrate, if the insulating layer is not to separate from its substrate. None the less, ceramic insulation has been used successfully in diesel engines–for example, the work reported by Walzer *et al.* (1985). In this turbocharged engine, 80 per cent of the combustion chamber surface was covered to an average depth of 3 mm by aluminium titanate or zirconium dioxide insulation. These measures led to a 13 per cent reduction in heat flow to the coolant, and a 5 per cent improvement in the urban cycle fuel economy (this was without re-optimisation of the cooling system).

In addition to describing insulating layers, complete ceramic components are described by Parker (1985). In this work, entire pistons have been made from reaction-bonded silicon nitride, and valves have been made from Syalon. Some additional advantages of ceramics are their high resistance to wear, and their low density (and thus lighter components). Consequently, ceramics are attractive for other components, such as valve guides, valve seats and tappets or other cam followers.

Reducing the heat transfer from the combustion chamber leads to several gains:

(a) An increase in the work produced during the expansion stroke.
(b) A reduction in the cooling system power requirements for the coolant pump and especially the air cooling fan.
(c) A higher exhaust gas temperature.
(d) Reduced ignition delay and hence reduced diesel knock.

A study of a low heat loss engine by Hay *et al.* (1986) suggested that a 30 per cent reduction in heat transfer would lead to a 3.6 per cent reduction in the specific fuel consumption at the rated speed and load. Furthermore, the 2 kW reduction in the cooling fan power would lead to an additional 2.7 per cent reduction in the fuel consumption, at the rated speed and load, and correspondingly greater percentage improvements at part load. Work by Wade *et al.* (1984) in a low heat loss engine suggests that the greatest gains in efficiency are at light loads and high speeds (figure 3.9).

In a naturally aspirated engine, the higher combustion chamber temperature will lead to a reduction in the volumetric efficiency, and this will offset some of the gains from the increased expansion work. The fall in volumetric efficiency is less significant in a turbocharged engine, not least since the higher exhaust temperature will lead to an increase in the work available from the turbine; this is illustrated in figure 3.10. The increased turbine work makes the turbo-compound engine advocated by Wallace *et al.* (1983) more attractive; in this system, a turbine is coupled to the engine crankshaft. Alternatively, the higher exhaust temperature will allow a smaller turbine back-pressure for the same work output. Work by Hoag *et al.* (1985) indicates that a 30 per cent reduction in heat transfer would lead to a 3.4 per cent fall in volumetric efficiency, a 70 K rise in the exhaust gas temperature, and

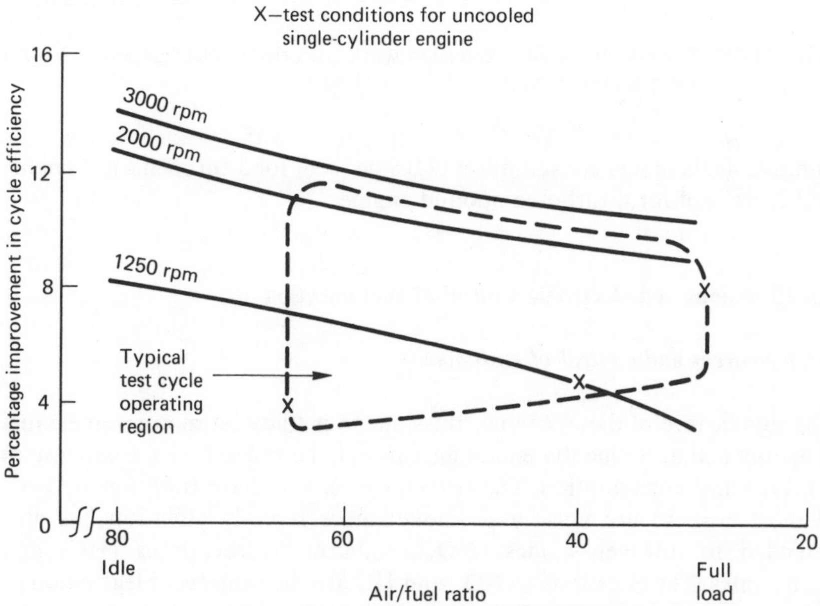

Figure 3.9 Calculated improvement in cycle efficiency for a fully insulated DI diesel engine relative to a baseline water-cooled DI diesel engine as a function of air/fuel ratio and engine speed. Bore 80 mm, stroke 88 mm, compression ratio 21:1, adapted from Wade *et al.* (1984)

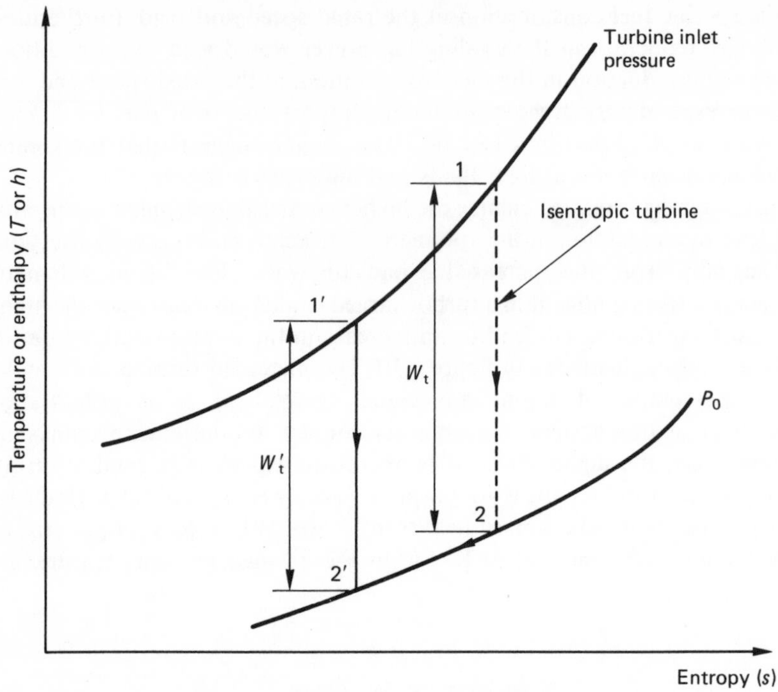

Figure 3.10 The rise in turbine expansion work with rising inlet temperature for a given pressure ratio

improvements in fuel consumption of 0.8 per cent for a turbocharged engine and 2 per cent for a turbo-compound engine.

3.5 Emissions and electronic control of fuel injection

3.5.1 Sources and control of emissions

The significance of diesel engine emissions in a study on motor vehicle fuel economy is that frequently emissions can only be reduced at the expense of a greater fuel consumption. The two emissions that have traditionally been of most concern are noise and smoke; more recently attention has also extended to nitrogen oxides (NO_x), unburnt hydrocarbons (HC) and particulates. The emissions of NO_x and HC are the subjects of legislation in Europe, the USA and Japan, and particulates legislation also exists in the USA. Carbon monoxide emissions need not be considered here, as a correctly regulated diesel engine always has negligible emissions of carbon monoxide, since the air/fuel ratio is always lean.

The noise from diesel engines is usually mostly attributable to the combustion noise, which originates from the high rates of pressure rise during the initial rapid combustion. A typical relationship between combustion noise and the peak rate of combustion is shown in figure 3.11. To understand the source of the rapid combustion, and how it can be controlled, it is necessary to consider the fuel injection and combustion processes.

As mentioned earlier, the fuel is injected into the combustion chamber towards the end of the compression stroke. The fuel evaporates from each droplet of fuel, and mixes with air to form a flammable mixture. However, combustion does not occur immediately, and during the delay period (between the start of injection and the start of combustion) a flammable mixture is being continuously formed. Consequently, when ignition occurs it does so at many sites, and there is very rapid combustion of the mixture formed during the delay period. This rapid combustion produces the characteristic diesel knock. Evidently the way to reduce the combustion noise is either to reduce the quantity of mixture prepared during the delay period, and this can be achieved by reducing the initial rate of injection, or more commonly to reduce the duration of the delay period. The delay period is reduced by having:

(a) A higher temperature, which increases both the rate of heat transfer and the chemical reaction rates.

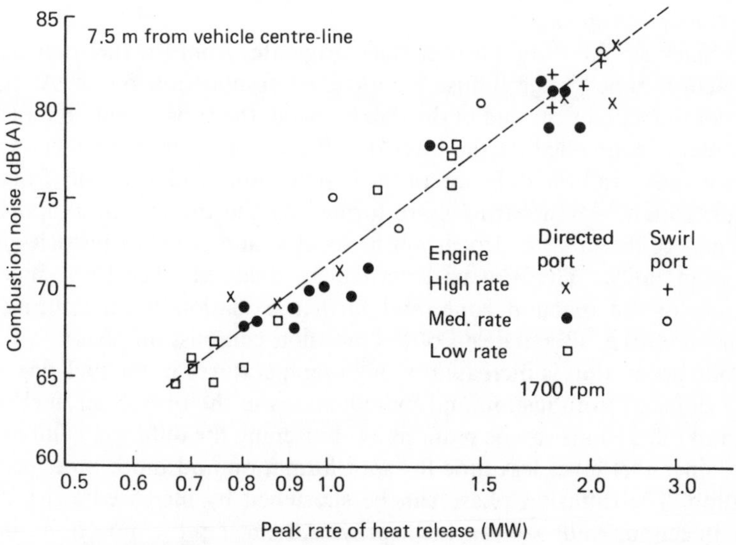

Figure 3.11 Relationship between combustion noise and peak rate of heat release, from Glikin (1985). [Reprinted by permission of the Council of the Institution of Mechanical Engineers]

(b) A higher pressure, which increases the rate of heat transfer.
(c) A fuel that spontaneously ignites more readily (a higher cetane number).

A higher cetane rating fuel is unlikely to be feasible, since for the reasons presented in section 3.2.2, the quality of diesel fuel is currently falling.

Higher temperatures occur in turbocharged engines and low heat loss engines, but as will be seen later, this also leads to higher NO_x emissions. A higher compression ratio would lead to both higher pressures and temperatures, but table 3.2 has already shown that this can lead to an increase in fuel consumption and smoke. An alternative approach is to retard the injection timing, so that injection occurs closer to the end of the compression stroke. This leads to an increase in the fuel consumption and other trade-offs that are discussed later in figure 3.12, for a fixed high load.

The trade-off between specific fuel consumption and injection timing for different injection rates is shown in figure 3.12. Ideally, combustion should occur instantaneously at top dead centre. In practice, combustion commences before top dead centre and continues afterwards. Advancing the start of injection (and thus combustion) increases the compression work, but the ensuing higher pressures and temperatures at top dead centre also increase the expansion work. However, if the injection timing is advanced too much, the increase in compression work will be greater than the increase in expansion work. Clearly, faster injection leads to more rapid combustion, and this results in less advanced injection timings. There is a rate of injection above which no further gains in fuel consumption occur, and the higher the swirl, the lower this injection rate.

The black smoke from diesel engines originates from the fuel-rich side of the reaction zone in the diffusion-controlled combustion phase. After the rapid combustion at the end of the delay period, the subsequent combustion of the fuel is controlled by the rates of diffusion of air into the fuel vapour and vice versa, and the diffusion of the combustion products away from the reaction zone. Carbon particles are formed by the thermal decomposition (cracking) of the large hydrocarbon molecules, and the soot particles form by agglomeration. The soot particles can be oxidised when they enter the lean side of the reaction zone, and further oxidation occurs during the expansion stroke, after the end of the diffusion combustion phase.

Smoke generation is increased by high temperatures in the fuel-rich zone during diffusion combustion and by reductions in the overall air/fuel ratio. The smoke emissions can be reduced by shortening the diffusion combustion phase, since this gives less time for soot formation and more time for soot oxidation. The diffusion phase can be shortened by increased swirl, more rapid injection, and a finer fuel spray. Advancing the injection timing also reduces the smoke emissions. The earlier injection leads to higher temperatures during the expansion stroke, and more time in which oxidation of the soot particles can occur. Unfortunately, advancing the injection timing

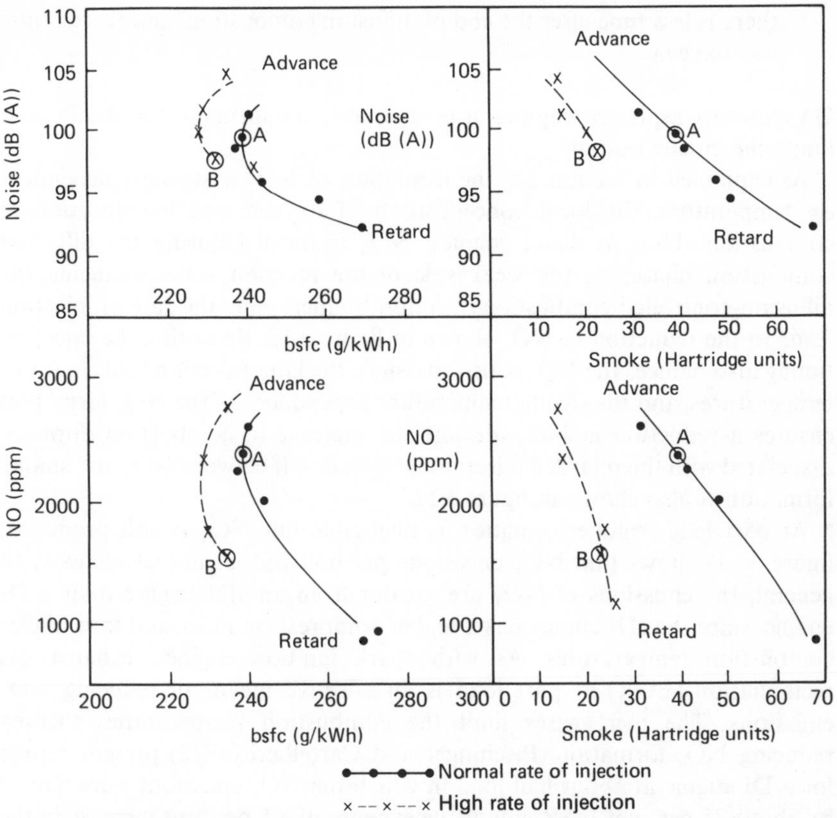

Figure 3.12 Trade-off curves between noise, smoke, NO_x and specific fuel consumption for different rates of injection and injection timing, from Glikin (1985). [Reprinted by permission of the Council of the Institution of Mechanical Engineers]

leads to an increase in noise. However, if the injection rate is increased and the timing is retarded, there can be an overall reduction in both noise and smoke. One such combination of points is shown as A and B in figure 3.12, and it can also be seen that the minimum specific fuel consumption has been reduced slightly, and that there is a significant reduction in nitrogen oxide emissions.

The formation of smoke is most strongly dependent on the engine load. As the load increases, more fuel is injected, and this increases the formation of smoke for three reasons:

(a) The duration of diffusion combustion increases.
(b) The combustion temperatures increase.
(c) Less oxidation of the soot occurs during the expansion stroke since

there is less time after the end of diffusion combustion, and there is also less oxygen.

On naturally aspirated engines, it is invariably the formation of smoke that limits the engine output.

As explained in section 2.4, the formation of NO_x is strongly dependent on temperature, the local concentration of oxygen and the duration of combustion. Thus in diesel engines, NO_x is formed during the diffusion combustion phase, on the weak side of the reaction zone. Reducing the diffusion-controlled combustion duration by increasing the rate of injection leads to the reduction in NO_x shown in figure 3.12. Retarding the injection timing also reduces the NO_x emissions, since the later injection leads to lower temperatures, and the strong temperature dependence of the NO_x formation ensures a reduction in NO_x, despite the increase in combustion duration associated with the retarded injection. The trade-off between NO_x and smoke formation is also shown in figure 3.12.

At part load, smoke formation is negligible but NO_x is still produced; figure 3.13a shows that NO_x emissions per unit output in fact increase. In general, the emissions of NO_x are greater from an IDI engine than a DI engine, since the IDI engine has a higher compression ratio, and thus higher combustion temperatures. As with spark ignition engines, exhaust gas recirculation (EGR) at part load is an effective means of reducing NO_x emissions. The inert gases limit the combustion temperatures, thereby reducing NO_x formation. Pischinger and Cartellieri (1972) present results for a DI engine at about half load in which the NO_x emissions were halved by about 25 per cent EGR, but at the expense of a 5 per cent increase in the fuel consumption.

Unburnt hydrocarbons (HC) in a properly regulated diesel engine come from two sources. Firstly, around the perimeter of the reaction zone there will be a mixture that is too lean to burn, and the longer the delay period, the greater the amount of HC emissions from this source. However, there is a delay period below which no further reductions in HC emissions are obtained. Under these conditions the HC emissions mostly originate from a second source, the fuel retained in the nozzle sac (the space between the nozzle seat and the spray holes) and the spray holes. Fuel from these sources can enter the combustion chamber late in the cycle, thereby producing HC emissions.

Figure 3.13b shows that HC emissions are worse for DI engines than IDI engines, especially at light load, when there is significant ignition delay in DI engines. Advancing the injection timing reduces HC emissions, but this would lead to increased NO_x (figure 3.13b) and noise. The HC emissions increase at part load, the delay period increases, and the quantity of mixture at the perimeter of the reaction zone that is too lean to burn increases. Finally, if a diesel engine is over-fuelled, there would be significant hydrocarbon emissions as well as black smoke.

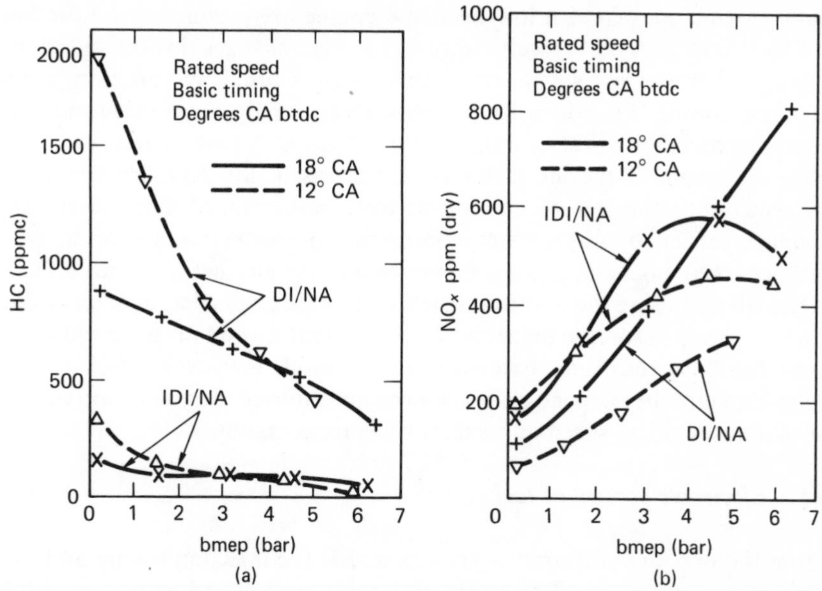

Figure 3.13 A comparison of emissions from naturally aspirated direct (DI) and indirect (IDI) injection diesel engines: (a) nitrogen oxide emissions, (b) hydrocarbon emissions. From Pischinger and Cartellieri (1972). [Reprinted with permission © 1972 Society of Automotive Engineers, Inc.]

The final class of emissions discussed here are particulates; these are any substance, apart from water, that can be collected by filtering diluted exhaust at a temperature of 325 K. Since the particulates are either soot or condensed hydrocarbons, any measure that reduces either the exhaust smoke or HC emissions should also reduce the level of particulates. If additional measures are needed, the particulates can be oxidised by a catalyst incorporated into the exhaust manifold, in the manner described by Enga *et al.* (1982). Some concern has also been expressed that the particulates may contain carcinogens, and a technique for identifying mutagens is presented by Seizinger *et al.* (1985).

Emissions from turbocharged engines and low heat loss engines are, in general, lower than from naturally aspirated engines; the exception is the increase in nitrogen oxide emissions. The higher combustion temperatures (and also pressure in the case of turbocharged engines) leads to a shorter ignition delay period, and a consequential reduction in the combustion noise and hydrocarbon emissions. The higher temperatures during the expansion stroke encourage the oxidation reactions, and the emissions of smoke and particulates both decrease. Nitrogen oxide (NO_x) emissions increase as a direct consequence of the higher combustion temperatures. Comprehensive data on emissions from a low heat loss engine are presented by Cole and Alkidas (1985). These results show that the changes in emissions are small, and much less significant than changes caused by variation of the injection timing. Emissions from turbocharged engines are discussed very comprehensively by Watson and Janota (1982), with particular attention to NO_x emissions. If the air/fuel ratio is increased, the NO_x emissions are reduced, since the lower combustion temperature more than offsets any increase caused by the greater concentration of oxygen. However, in a turbocharged engine at a given power rating, the air/fuel ratio can only be increased by raising the boost pressure. Unless the engine has an inter-cooler after the compressor, the higher air inlet temperature and pressure will result in higher NO_x emissions. Evidently the trade-offs with NO_x emissions are complex, since they depend on boost pressure, inter-cooling, injection timing, injection rate, load, speed and exhaust gas recirculation.

3.5.2 Electronic control of fuel injection

From the previous section it is very clear that the injection timing and rate both have a profound effect on the fuel consumption and emissions. While the injection system has to remain hydro-mechanical, because of the high injection pressures, electronic control of the injection timing and fuel quantity enables the engine performance to be optimised for variously low emissions, high output and high economy. An electronic control system described by Glikin (1985) is shown in figure 3.14. The signals from the control unit operate through hydraulic servos in otherwise conventional injector pumps. The optimum injection timing and fuel quantity are controlled by the micro-processor in response to several inputs: driver demand, engine speed, turbocharger boost pressure, air inlet temperature and engine coolant temperature. A further refinement used by Toyota is to have an optical ignition timing sensor for providing feedback; this is described by Yamaguchi (1986b). The ignition sensor enables changes in ignition delay to be detected, and the control unit can then re-optimise the injection timing accordingly. This system thus allows for changes in the quality of the fuel and the condition of the engine.

A more radical appoach is to use a unit injector with electronic control. Unit injectors incorporate the injector pump and nozzle into a single unit; by combining these two elements, many problems associated with the interconnecting pipes are eliminated. Figure 3.15 shows the performance of a conventional system, in which the compressibility of the fuel and elasticity of the connecting pipe have significant effects. The pumping pressure does not rise instantaneously, and time is also needed for the pressure to propagate to the injector. Injection does not start until the nozzle opening pressure has been reached, and injection continues for some time after the end of pumping. The propagation delay is of approximately constant duration, and this

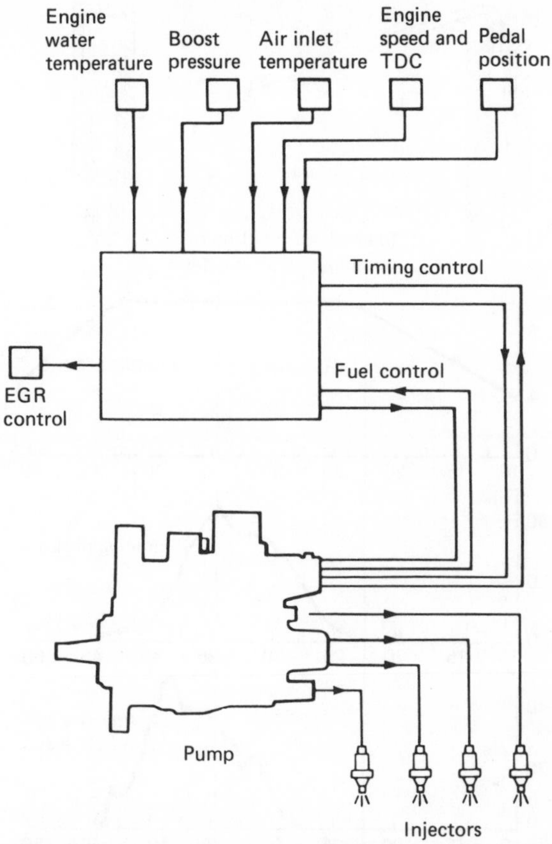

Figure 3.14 Schematic arrangement of electronic fuel injection control, Glikin (1985). [Reprinted by permission of the Council of the Institution of Mechanical Engineers]

occupies a larger cam angle at higher speeds. The ratio of the mean pumping rate to the mean injection rate (the spread-over ratio) also increases with higher engine speeds. The injection duration increases at high speeds in terms of cam or crankshaft angle, and it can be over twice the angular duration of low speeds. Since the swirl speed increases directly with engine speed, there will either be too much swirl at high speeds or insufficient swirl at low speeds. (Shorter duration injection requires either more holes in the nozzle or a higher rate of swirl to ensure complete mixing of the fuel and air.)

With a unit injector, the delay between the start of pumping and the start of injection is much reduced, and the spread-over ratio remains much closer to unity. However, a fast acting electrically operated hydraulic control valve is needed, and the Colenoid described by Seilly (1981) has been developed for this purpose. Indeed, the response is so rapid that pilot injection can be

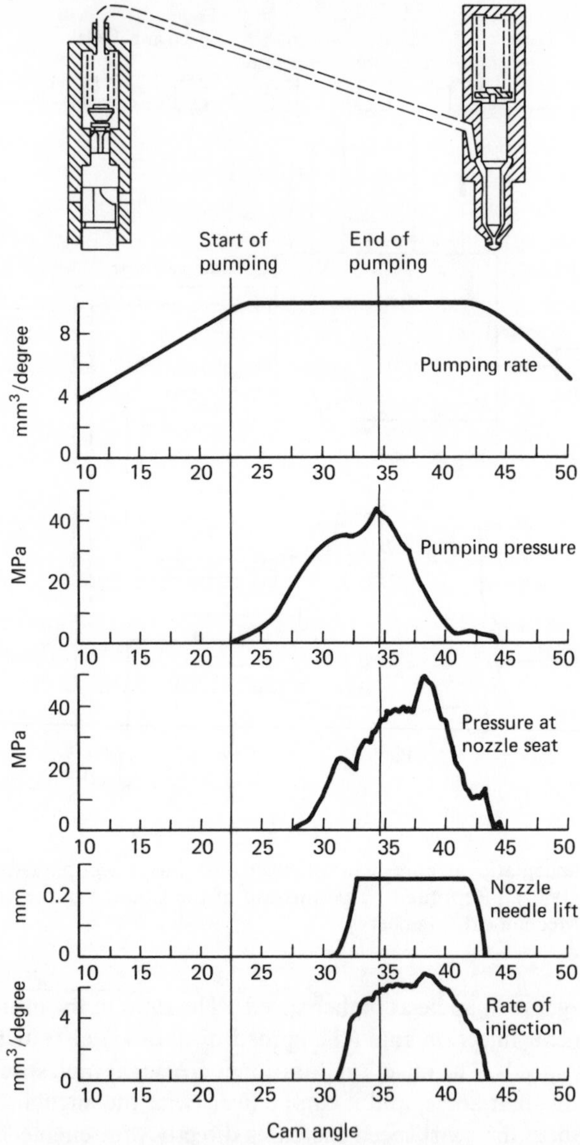

Figure 3.15 Injection diagrams for pump–pipe–nozzle system, Glikin (1985), [Reprinted by permission of the Council of the Institution of Mechanical Engineers]

achieved. By injecting the fuel in two stages, the bulk of fuel delivery can be delayed until after the start of combustion. This reduces the significance of the ignition delay period, with consequential improvement in the hydrocarbon and combustion noise emissions.

3.6 Conclusions

Whenever the performance of diesel and petrol engined vehicles is being compared, due account ought to be taken of the different energy content of the fuels (table 3.1) and of the different fuel costs. None the less, diesel engines have a higher maximum efficiency than petrol engines for four reasons: the compression ratio is higher, initially only air is present during compression, the fuel/air mixture is always weak of stoichiometric, and the operating speed is usually slower. Furthermore, the fall in part load efficiency is less pronounced since there is no throttling, the air fuel mixtures become weaker, and the combustion duration reduces. Any comparison also needs to consider the output per unit swept volume and the output per unit weight. Spark ignition engines can quite easily obtain outputs of 40 kW/litre, while a naturally aspirated direct injection diesel engine might only achieve 20 kW/litre. Turbocharging diesel engines in automotive applications frequently increases the output by about 40 per cent to 28 kW/litre. In addition, the weight of diesel engines per unit swept volume is usually greater than that of spark ignition engines. In comparably powered passenger cars, the use of a diesel engine may introduce a 6 per cent weight penalty. It is shown later (in section 6.3) that a 6 per cent weight penalty might incur a fuel consumption penalty of 3–5 per cent. Frequently, the diesel engined vehicle is the least powerful model in a vehicle range: manufacturers assume that customers who are looking for good fuel economy and reliability will be prepared to accept a lower performance.

The differences in specific fuel consumption between spark ignition engines and indirect injection diesel engines are small (figure 3.2), apart from at light load. The only way to assess the fuel economy benefits of a diesel engined vehicle is by using the approach developed later in section 4.2.1, or preferably a computer model of the type described in section 4.6.

As the speed range of direct injection diesel engines has increased (now 4500 rpm), the size at which they offer the same output as indirect injection engines has reduced to about 500 cm^3 per cylinder. Figure 3.4 shows that the direct injection diesel engine has at least a 10 per cent lower fuel consumption than the indirect injection engine, over almost the entire load speed range. The explanation for the fuel economy improvement, (as explained in section 3.3), stems from the lower heat transfer losses, and the reduced frictional losses as a consequence of the lower compression ratio.

Both direct and indirect injection engines benefit from turbocharging, as the increase in specific output is accompanied by a reduction in fuel consumption of a few per cent. Similarly, the use of low heat rejection designs (section 3.4.2) also offers a further reduction in fuel consumption, especially when used in conjunction with a turbocharger. Performance maps do not appear to be available for low heat rejection engines, so a thorough comparison with conventionally cooled engines is difficult. However, the reduced power requirement of the cooling system will in itself lead to useful savings in fuel consumption at part load. In heavy-duty truck engines, the use of low heat rejection turbo-compound engines with variable geometry turbochargers may be justified.

While emissions from diesel engines are inherently lower than those from spark ignition engines, they cannot be ignored. In section 3.5.1, the sources of the emissions (noise, smoke, particulates, unburnt hydrocarbons and nitrogen oxides) have been identified; their dependence on injection timing and rate is illustrated by figure 3.12. The effect of load and injection timing on emissions of unburnt hydrocarbon and nitrogen oxides is shown in figure 3.13; unburnt hydrocarbon emissions from direct injection engines are a cause for concern. The trade-offs between power output, efficiency and the different types of emission mean that careful choice of injection timing is needed. Furthermore, to ensure that the optimum injection timings are obtained, it is necessary to have an electronic control system, of the type described in section 3.5.2.

In conclusion, diesel engines offer very good fuel economy but with a lower specific output than spark ignition engines. Further improvements are possible with the advent of low heat loss engines and the use of electronic fuel injection control, to provide the optimum fuel economy within acceptable emissions limits.

3.7 Discussion points

(1) Why is the efficiency of a diesel engine greater than that of a spark ignition engine? Why does the efficiency of a diesel engine fall less rapidly than the part load efficiency of a spark ignition engine?
(2) What are the fundamental differences between direct injection (DI) and indirect injection (IDI) diesel engines, and why is the DI engine more efficient? Why is the combustion speed a limitation only in small-capacity engines?
(3) What are the advantages that turbocharging and reducing heat transfer offer to the performance of diesel engines?
(4) How does injection timing affect the delay period, the rapid combustion and the diffusion controlled combustion phase?
(5) How are combustion noise, fuel consumption, emissions of unburnt hydrocarbons and nitrogen oxides affected by (a) the injection timing, (b) the rate of injection?
(6) What are the parameters that affect emissions of nitrogen oxides, and what is their effect?

4 Transmission Systems

4.1 Introduction

The transmission has to transfer power from the engine to the road wheels in order to propel the vehicle. The essential components are a clutch or hydraulic coupling (to isolate the engine from the transmission), a gearbox (to allow the vehicle speed range to be greater than the engine speed range and to provide load matching) and a differential (to allow relative rotation of the driven wheels during turns). The differential can be remote from the gearbox and form part of the rear axle. In front-engined front wheel drive vehicles and in rear-engined real wheel drive vehicles, the gearbox and differential can be combined: an arrangement that is sometimes referred to as a transaxle. In high-performance vehicles with the engine at the front, the gearbox is sometimes incorporated into the back axle in order to improve the weight distribution.

The basic requirements for the transmission are:

(a) The overall gearing should be such that the maximum vehicle speed can be obtained from the available power.
(b) The vehicle should be able to start, fully laden, on a steep gradient, typically 1 in 3 (or 33 per cent).

Since the maximum speed of a vehicle is usually obtained when the gearbox is in direct drive (no power transmission through any gears) it is the reduction ratio in the back axle or final drive that has to be chosen so that the maximum power occurs at the maximum vehicle speed. In practice the reduction ratio may be slightly greater, since this has a small effect on the maximum vehicle speed, but enables the maximum speed to be maintained against a gradient or headwind.

The hill-starting requirement will define a torque requirement into the final drive, and since this is invariably greater than the maximum torque available from the engine, a reduction gearing is needed. As the reduction ratio for starting may be 4:1 in a car and 12:1 for a commercial vehicle, then intermediate gear ratios are necessary if there is to be a steady and effective increase in vehicle speed, and the vehicle is to be 'driveable'.

While the speed range of engines has increased, the speed range of vehicles has increased even more. In the case of cars this led to three-speed (that is, gearing ratios) gearboxes being replaced by four-speed gearboxes. With automatic transmissions, three-speed gearboxes have been usual because of the hydraulic torque converter. The torque converter (described more fully in section 4.4) combines two separate roles: at idling speed the torque converter isolates the engine from the transmission, and secondly as the input speed is increased, an output torque is produced as a consequence of the 'hydraulic gearing'. The main disadvantage of torque converters is that their efficiency is invariably less than 90 per cent, and often significantly less.

Defining the gearing ratios between the engine and driven wheels to ensure adequate hill starting, and maximum vehicle speed, does not lead to the optimum fuel economy. When a vehicle is cruising at a speed below full throttle, the engine will be running at a part load condition. The engine efficiency can then invariably be improved by obtaining the same power at a reduced speed but at a higher load. However, this requires the use of an overdrive unit or an extra gear ratio. The fuel savings from an overdrive ratio are primarily due to reducing the power lost through mechanical inefficiency in the engine. Additional advantages of overdrive ratios are the reductions in engine wear and noise.

A special class of automatic transmission that will have increasing significance is the continuously variable transmission (CVT). As the name suggests, the discrete steps associated with conventional gearbox systems are eliminated, and the range of gearing ratios (the span) is also much wider.

When the performance of a powertrain is being analysed, due account should be made for losses in the transmission. Since the efficiency of gear systems is inherently high, accurate efficiency measurements are difficult to make. Furthermore, the efficiency of a particular element will be a function of the load and speed. The frictional losses in the bearings and seals will be essentially a function of speed, while the frictional losses between gear teeth will also depend on the load. For simplicity and expediency the following transmission efficiencies will be assumed:

gearbox–direct drive (top gear)	98%
gearbox–indirect drive lower gears or over drive	95%
rear axle	95%
CVT	90%

torque converter is a function of the operating point (see, for example, figure 4.8)

The efficiency of any gearbox is influenced by its lubricant, and Padmore (1977) reviews the effect of lubricants on fuel economy, along with comprehensive efficiency data for a test on a rear axle. The difficulties in assessing the effects of the engine lubricant on the vehicle fuel economy also arise in establishing the effects of transmission lubricants on vehicle fuel economy. Data from tests on rear axles by Goodwin and Haviland (1978) again show that the most significant improvements occur with urban driving. The change from a 90 to 75 W grade oil gave fuel consumption improvements ranging between 1.0 and 3.1 per cent.

When transmission efficiency data are available, a computer model is the appropriate way to amalgamate the engine and transmission efficiency data, in order to predict vehicle fuel economy. The efficiency of the rear axle is comparatively low (for the number of gears meshing) because of the nature of the contact between the teeth in hypoid gears. Hypoid gears are a form of spiral bevel gear used in cars, since this form enables the axis of the driveshaft to be below the axis of the wheels. However, this means that the contact between the gear teeth does not have a pure rolling motion, and the presence of a sliding velocity leads to frictional losses; the magnitude of the sliding velocity is a function of the offset between the gear axes.

The following sections first develop the general principles of transmission matching, through an extended example that considers conventional gearboxes, overdrive ratios, continuously variable transmissions and the use of a diesel engine. Then follows a discussion of the mechanical details of the different transmission elements; further details can be found in many books–for example, Newton *et al.* (1983). Finally, some different approaches to powertrain optimisation are discussed.

4.2 Powertrain matching–an extended example

4.2.1 General principles

Since the principles in matching the gearbox and engine are essentially the same for any vehicle, it will be sufficient to discuss just one vehicle. The example used here is a car, with the following specification:

rolling resistance, R	225 N
drag coefficient C_d	0.33
frontal area, A	2.25 m^2
required top speed	160 km/h
mass	925 kg

The tractive force (T) is shown as a function of speed (v) for this vehicle in figure 5.12. At a speed of 160 km/h a power of 49 kW is required (power, $\dot{W}_b = T \times v$). For the sake of this discussion a manual gearbox will be assumed, but the same general principles will apply for automatic transmissions. The term 'top gear' will refer here to either the third gear in an automatic gearbox or the fourth gear in a manual gearbox. Similarly, the term 'overdrive' will refer here to either the fourth gear in an automatic gearbox or the fifth gear in a manual gearbox.

To travel at 160 km/h, a power of 49 kW is needed at the wheels ($\dot{W}_w$); to find the necessary engine power ($\dot{W}_b$), divide by the product of all the transmission efficiencies ($\Pi\eta$).

$$\dot{W}_b = \frac{\dot{W}_w}{\Pi\eta} \tag{4.1}$$

Efficiencies: rear axle 95%
top gear 98%

Thus

$$\dot{W}_b = \frac{49}{0.95 \times 0.98} = 52.6 \text{ kW}$$

Suppose a four stroke spark ignition engine is to be used that has the characteristics that are defined by the map in figure 4.1. The contours show

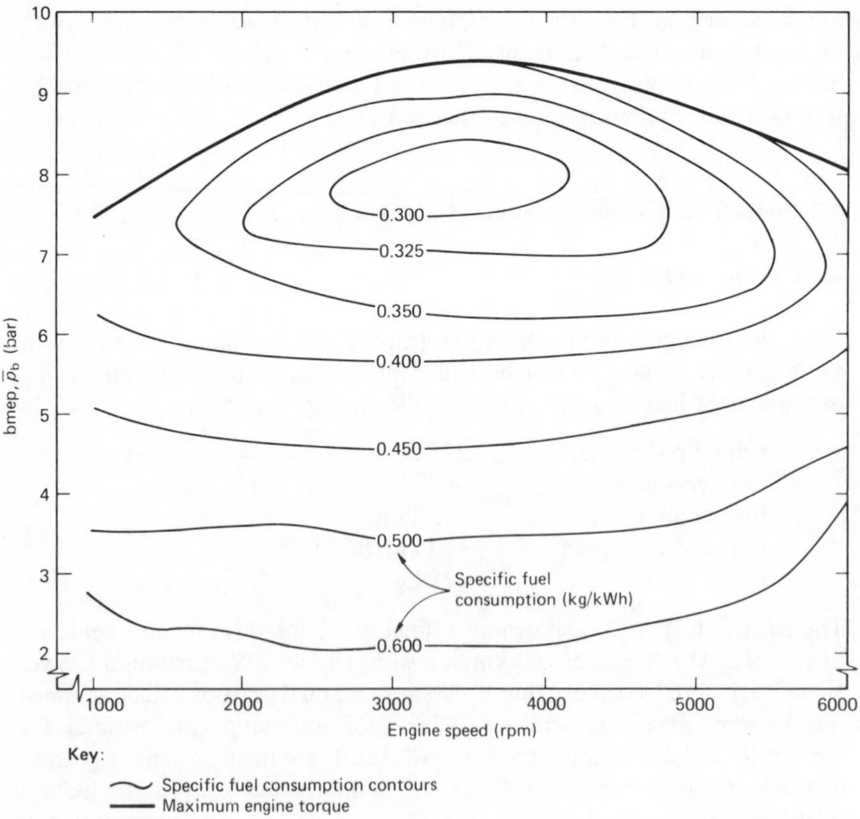

Figure 4.1 Spark ignition engine map

lines of constant specific fuel consumption, which is proportional to the reciprocal of efficiency. If the top speed of the vehicle (160 km/h) is to coincide with the maximum engine speed (6000 rpm), then the overall gearing ratio is such as to give 26.7 km/h per 1000 rpm. At 6000 rpm the brake mean effective pressure (bmep) is 8.1 bar; if the power required is 52.6 kW, the swept volume necessary can be found by rearranging equation 2.2:

$$V_s = \frac{\dot{W}_b}{p_b \times N'} = \frac{52.6 \times 10^3}{8.1 \times 10^5 \times 6000/120} = 1300 \text{ cm}^3 \qquad (4.2)$$

Now that the swept volume has been determined, the bmep axis on figure 4.1 can be recalibrated as a torque, T.

$$\text{Power, } \dot{W}_b = p_b \times V_s \times N' \equiv T \times \omega \qquad \text{(equation 2.2)}$$

$$\text{For a four stroke engine, } \omega = 4\pi \times N'$$

$$T = \frac{p_b \times V_s}{4\pi} (\text{Nm}) \qquad (4.3)$$

Since the gearing ratios and efficiencies have been defined such that the maximum power of the engine corresponds to 160 km/h, the total tractive resistance curve (the propulsive force as a function of speed in figure 5.12) can be scaled to give the road load curve (the engine torque required for propulsion as a function of engine speed) which is shown on figure 4.2. This scaling automatically incorporates the transmission efficiencies, since they were used in defining the maximum power requirement of the engine.

Also identified on figure 4.2 are the points on the road load curve that correspond to speeds of 90 and 120 km/h. The difference in height between the road load curve and the maximum torque of the engine represents the torque that is available for acceleration and overcoming head winds or gradients. In the case of 120 km/h there is a balance of 41.8 N m. The torque (T) can be converted into a tractive effort, since the overall efficiency and gearing ratios are known.

$$\text{gearing ratio } (gr) = 26.67 \text{ km/h per 1000 rpm}$$

$$\equiv 26.67/60 = 0.444 \text{ m/rev.} \qquad (4.4)$$

$$(gr) = 0.444/2\pi = 0.07074 \text{ m/radian}$$

The residual tractive force available is

$$\frac{T}{gr} \times \eta_{\text{axle}} \times \eta_{\text{gearbox}}$$

$$= \frac{41.8}{0.07074} \times 0.95 \times 0.98 = 550\text{N}$$

Since the vehicle mass is 925 kg its weight is 9074 N, thus 120 km/h can be

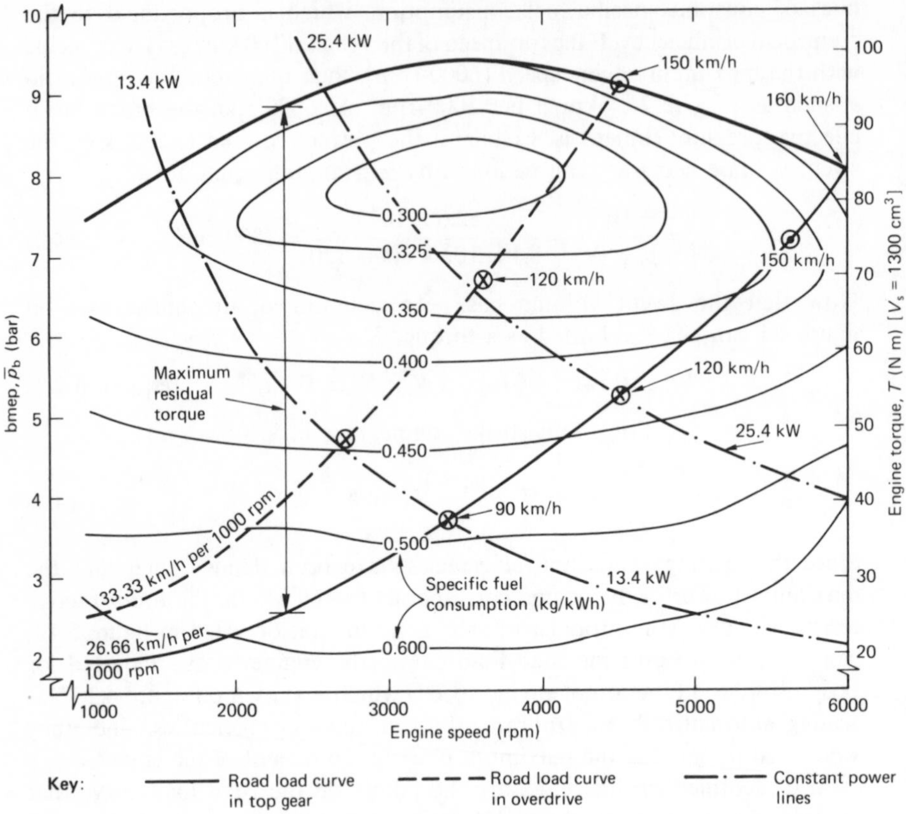

Figure 4.2 Road load curves and constant power lines added to figure 4.1

maintained up a gradient of 550/9074 = 6.0 per cent. If this gradient is exceeded then the vehicle will slow down until there is sufficient torque available to maintain a constant speed. As the speed reduces, the torque required for steady level running is given by the road load curve, and the torque available is determined from the engine torque curve. The rate at which this difference increases as speed reduces is referred to as the torque back-up. A high torque back-up gives a vehicle good driveability, since the speed reduction when gradients are met is minimised, and the need for gear changing is also minimised. The maximum residual torque available in top gear for hill climbing occurs at 2200 rpm (which corresponds to 59 km/h); if the speed reduces beyond this point, the torque difference decreases, and assuming the gradient remains unchanged, the engine would soon stall. In practice, a gear change would be made long before this point is met, since a driver would normally attempt to maintain speed by operating the engine close to the maximum power point of the engine.

At 120 km/h, the residual tractive effort of 550 N would enable the speed to be maintained against a 54 km/h headwind; this is shown by example 5.2 in section 5.7.

By interpolation on figure 4.2, the specific fuel consumption of the engine can be estimated as 0.43 kg/kWh at 120 km/h and 0.49 kg/kWh at 90 km/h. The power requirement at each operating point can be found from the product of torque and speed. An alternative approach is to use the tractive resistance curve shown in figure 5.12, making allowance for the transmission efficiencies as in equation 4.1, to calculate the power requirement.

Since the specific fuel consumption is also known, it is possible to calculate the steady-state fuel economy at each speed; these results are all summarised in table 4.1.

Figure 4.2 shows quite clearly that none of these operating points is close to the area of the highest engine efficiency. Since power is the product of torque and speed, lines of constant power appear as hyperbolae on figure 4.2. The operating point for minimum fuel consumption is where these constant power hyperbolae just touch the surface defined by the specific fuel consumption contours. This optimal economy operating line has been added in figure 4.3. However, if a vehicle with a conventional manual or automatic gearbox was designed to operate at these points, there would be so little torque back-up that the vehicle would be undriveable.

4.2.2 Overdrive gear ratios

Suppose a gearbox with an overdrive ratio of 0.8:1 and an efficiency of 95 per cent was to be fitted; from this information it is possible to draw the additional road load line in figure 4.2 as follows. The overall gearing ratio is now 33.33 km/h per 1000 rpm, and this enables the x-axis to be scaled directly. When scaling the y-axis, due account must be taken of the reduced efficiency of the overdrive ratio (95 per cent) as compared with the direct

Table 4.1 Summary of top gear performance figures for the vehicle defined by figure 4.2; all values are in terms of the engine output

Vehicle speed v (km/h)	Engine speed (rpm)	ω (rad/s)	Torque T (N m)	Power $\dot{W}_b$ $T \times \omega$ (kW)	sfc (kg/kWh)	Fuel economy $v/(\text{sfc} \times \dot{W}_b)$ $v/(\text{sfc} \times \dot{W}_b)$
90	3375	353.4	38	13.4	0.49	13.7
120	4500	471.2	54	25.4	0.43	11.0
150	5625	589.0	74.5	43.9	0.375	9.1

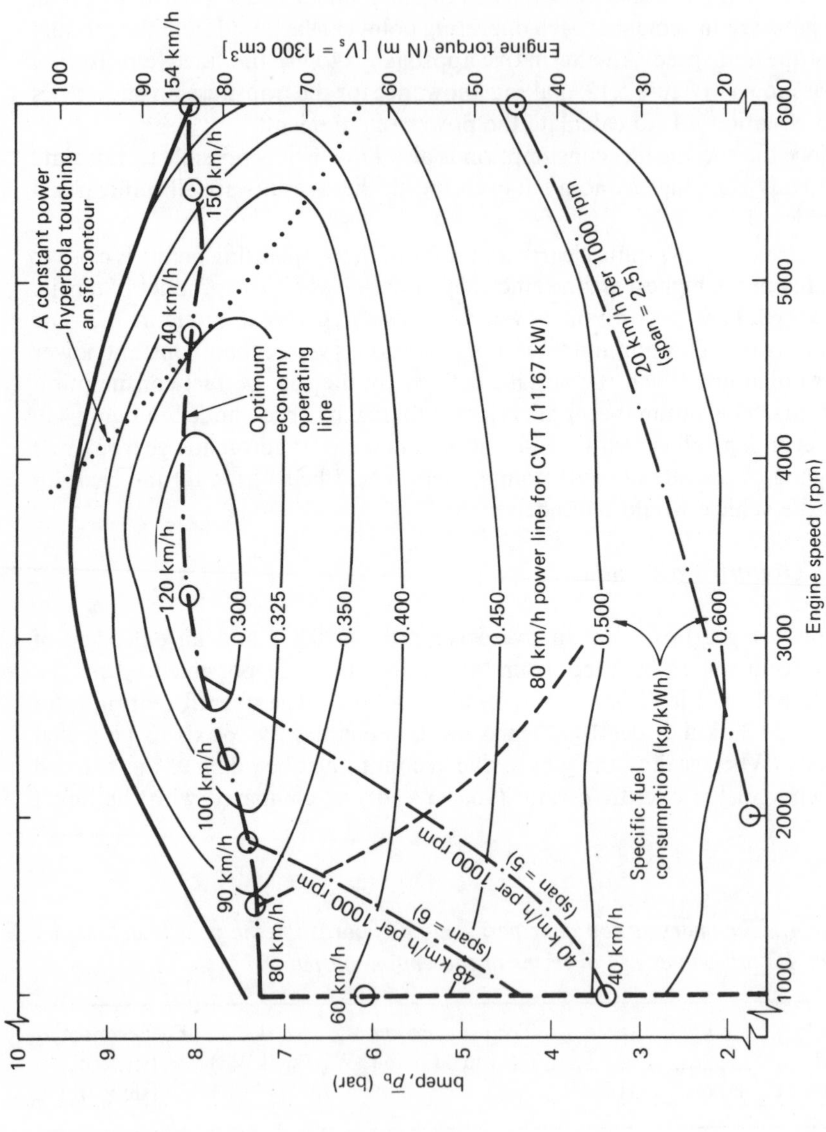

Figure 4.3 Locus for operation of a continuously variable transmission (CVT)

drive (98 per cent). From equation 4.1:

$$\text{Torque} = \text{tractive force} \times \text{gearing ratio} \times \frac{0.98}{0.95}$$

This reduction in efficiency also accounts for the slight increase in power requirements to 26.2 kW at 120 km/h and 13.8 kW at 90 km/h.

Figure 4.2 shows that the overdrive road load curve intersects the maximum engine torque curve at 4500 rpm. Since the gearing ratio is 33.33 km/h per 100 rpm, the top speed in overdrive will be 150 km/h; this is quite a small reduction in top speed from when top gear is used.

At 120 km/h the residual engine torque available is 29.2 N m (compared with 41.8 N m in top), and taking into account the transmission efficiency, this implies an ability to overcome 4.2 per cent gradients and 40 km/h headwinds. As before, the specific fuel consumption can be found at the operating points by interpolation on figure 4.2, and the fuel consumption can be calculated. Due account must be taken of the increased engine power output requirement that is caused by the reduced transmission system efficiency in overdrive; the results are summarized in table 4.2.

The use of an overdrive gear ratio can be seen to give a 24.5 per cent improvement in fuel economy at 120 km/h, and a 7.3 per cent improvement in fuel economy at 90 km/h in this particular case. At 150 km/h the fuel consumption is worse in overdrive, since the engine is operating at full throttle with a rich air/fuel mixture, and a correspondingly high specific fuel consumption. As mentioned before, the additional benefits of using an overdrive ratio are the reduced engine noise and wear.

4.2.3 Continuously variable transmissions

It has already been stated that the optimal efficiency for a given power output is where the constant power line just touches a specific fuel consumption

Table 4.2 Summary of overdrive performance figures for the vehicle defined by figure 4.2; all values are in terms of the engine output

Vehicle speed v (km/h)	Power $\dot{W}_b$ (kW)	sfc (kg/kWh)	Fuel economy $v/(\text{sfc} \times \dot{W}_b)$ (km/kg)	Improvement in fuel economy compared with top gear (%)
90	13.8	0.445	14.7	7.3
120	26.2	0.335	13.7	24.5
150	45.2	0.420	7.9	– 13.2

contour. The locus of these points defines the optimum operating line for a vehicle, and this is shown in figure 4.3. The only way this locus can be followed is by means of a Continuously Variable Transmission (CVT).

Assuming an efficiency of 90 per cent for the CVT and 95 per cent for the final drive, the power that is required to propel a vehicle with a road load curve defined by figure 5.12 can be calculated. The operating points for the CVT can then be found by interpolation on figure 4.3. However, as this engine cannot run below 1000 rpm, at speeds of about 70 km/h and lower, the engine has to operate at part throttle.

At 120 km/h it may appear that the residual torque available is only 14.6 N m, but this is not the case with a CVT. If more power is required at a given speed, the reduction ratio in the CVT increases, so that the engine speed can rise, and the operating point moves to the right on the locus. At 120 km/h the overall gearing ratio gives 38.4 km/h per 1000 rpm, while to obtain maximum torque at this road speed maximum power is needed (6000 rpm), and the gearing ratio would reduce to 20 km/h per 1000 rpm. To calculate the residual torque at this speed it is simplest to refer to the tractive effort at the road wheels.

Tractive effort required for 120 km/h: 722 N

Maximum power available at the engine: 52.6 kW

Maximum power at the wheels is found by including the transmission efficiencies: $52.6 \times 0.95 \times 0.9 = 45.0$ kW (4.5)

$$\text{The maximum tractive effort} = \frac{\text{Max. power}}{\text{Speed}}$$

$$= \frac{45 \times 10^3}{120 \times 10^3/(60 \times 60)} = 135 \text{ N}$$

Subtracting the tractive effort required for 120 km/h on level ground with no head wind gives: $1350 - 722 = 628$ N, thus 628 N is available for overcoming head winds or gradients when the gearing ratio is 20 km/h per 1000 rpm, which implies an ability to climb 6.9 per cent $(628/(925 \times 9.81))$ gradients. When compared with the performance in top gear of a fixed ratio gearbox, this is only a slight improvement since the CVT efficiency is lower and the difference in the gearing ratios is only slight. Figure 4.3 also shows that the top speed is reduced by 8 km/h because of the inefficiency of the CVT.

The main attraction of a CVT is the potential for improvements to the vehicle fuel economy. As before, the power requirement at each speed can be calculated, and the specific fuel consumption can be found from interpolation on figure 4.3; the results are summarised in table 4.3.

For convenience, the fuel economy figures, along with an indication of the gradient climbing ability and maximum velocity, are summarised in table 4.4 for the three difference transmission systems that have been discussed.

Table 4.3 Summary of performance for a vehicle defined by figure 4.3 with a CVT; all values are in terms of the engine output

Vehicle speed v (km/h)	Power $\dot{W}_b$ (kW)	sfc (kg/kWh)	Fuel economy $v/(\text{sfc} \times \dot{W}_b)$ (km/kg)
90	14.6	0.332	18.6
120	27.7	0.290	14.9
150	47.8	0.420	7.5

In summary, using an overdrive or CVT improves the vehicle fuel economy at all but the highest speeds, with the CVT providing the largest gains. At the highest speeds, the lower efficiency of an overdrive ratio or a CVT leads to the engine operating in a less efficient regime, thereby leading to a reduction in vehicle fuel economy.

The results in table 4.4 should be treated with caution since the engine map that has been used is typical but arbitrary, and constant values of efficiency have been assumed for the transmission elements. In practice, the map from a specific engine would be combined with a transmission system that should have efficiency recorded as a function of speed, load and reduction ratio. At this level, the calculations are most effectively performed by a computer; the logic and approach are, of course, the same as adopted here.

Table 4.4 Comparison of vehicle performance for different transmission systems

Vehicle speed (km/h)	Fuel economy		CVT (km/kg)
	Top gear	Overdrive	
90	13.7	14.7	18.6
120	11.0	13.7	14.9
150	9.1	7.9	7.5
Gradient (%) that can be ascended at 120 km/h	6.0	4.2	6.9
Maximum speed (km/h)	160.0	150.0	154.0

4.2.4 Gearbox span

The span of a gearbox is the ratio between the highest and lowest reduction ratios. In the introduction it was stated that the ability to start ascending a gradient (say 33 per cent) determined the lowest gear ratio. This requirement will be discussed for a car with a manual gearbox; in the case of an automatic gearbox, the torque converter will reduce the size of the reduction ratio needed and it will suffice here to quote some typical results for comparison.

Figure 4.2 shows that the engine produces a maximum torque of 99 N m at 3500 rpm; it will be assumed here that the clutch can transmit this torque with whatever slip is necessary, since initially the gearbox and final drive will be stationary. Thus if

Vehicle mass	925 kg
Vehicle weight	$925 \times g = 9074$ N
then, Tractive force necessary on a 33 per cent gradient	$= 0.33 \times 9074 +$ rolling resistance N
	$= 2994 + 225 = 3219$ N
Gearing ratio (equation 4.4), *gr*	$= \dfrac{99}{3219} 0.95 \times 0.98$
	$= 0.0286$ m/radian

This compares with 0.07074 m/radian in top gear, implying a reduction ratio of 2.47:1 for the lowest gear, since top gear is invariably a direct drive. In practice, the reduction ratio between first and fourth gears will be greater to reduce the peak demand on the clutch, and to provide some torque in reserve. If the starting requirement was based on an engine torque of 84 N m, then figure 4.2 shows that the clutch could be fully engaged at 1500 rpm as opposed to 3500 rpm.

Some typical gearbox ratios are given in table 4.5, along with the step-up ratios.

It may seem strange that the step-up ratios are not constant, since this would have given a geometric progression of gear ratios. A geometric progression would theoretically allow optimum acceleration by using the same position of the torque curve in each gear. However, in each case here the step-up ratio into top gear is smaller than the other step-up ratios. This is because the third gear of a manual gearbox (or second for an automatic gearbox) is used most frequently after top gear, for overcoming gradients or acceleration, and the step-up ratio associated with a geometric progression would be greater than optimum. An added advantage is the improved fuel economy in third gear when there is a small step-up ratio into top gear.

Since all the gear ratios have now been defined, it is possible to compute

Table 4.5 Summary of manual and automatic gearbox ratios

4-speed manual gearbox			3-speed automatic gearbox		
Gear	Reduction ratio	Step-up ratio	Gear	Reduction ratio	Step-up ratio
1	3.36	1.6	1	2.39	1.65
2	2.10	1.5	2	1.45	1.45
3	1.40	1.4	3	1.00	
4	1.00				

the tractive force available for each gear ratio as a function of vehicle speed. This is shown in figure 4.4, along with the tractive force that would be available with a CVT. Other than for first gear, the CVT tractive force curve lies below the maxima of the individual gear ratios; this is because the CVT efficiency is assumed to be 90 per cent. Another limitation with a CVT is the span (the range of ratios) that can be produced; in the case of the Perbury CVT the span is about 5. Since the largest reduction ratio is determined by the hill-starting requirement, the finite span of a CVT will then determine what the smallest reduction ratio will be. That is to say, the finite span limits the extent of the overdrive gearing that is available. In figure 4.3 the starting requirement is for an overall gearing ratio of 8 km/h per 1000 rpm. Consider this vehicle being driven at 120 km/h on level ground with no headwind, with the driver wishing to reduce speed to 80 km/h. Initially the throttle would remain fully open and, as the vehicle slowed-up, the engine would slow up faster as the overall gearing ratio reduced. However, if the span of this gearbox was 5, then at 112 km/h the gearing ratio would become its maximum of 40 km/h per 1000 rpm. Any further reduction in the vehicle speed would then be accomplished by a corresponding reduction in the engine speed, and the throttle closing as the operating point followed the 40 km/h per 1000 rpm road load curve to the 80 km/h operating point.

Referring again to figure 4.3, it can be seen that the full benefits of a CVT are not attainable with a span of either 5 or 6; in this particular case a span of 8 is necessary to follow the optimum locus. The limited span will also affect the fuel economy; for example, if the span is 6, then optimum fuel economy cannot be attained in the speed range of 48–90 km/h. The effect on fuel economy at 80 km/h is shown as a function of span in table 4.6.

Table 4.6 shows that if a CVT is to offer significant fuel economy improvements, it needs to be able to operate over a wide span of gear ratios; this is partially to offset the inherently lower efficiency.

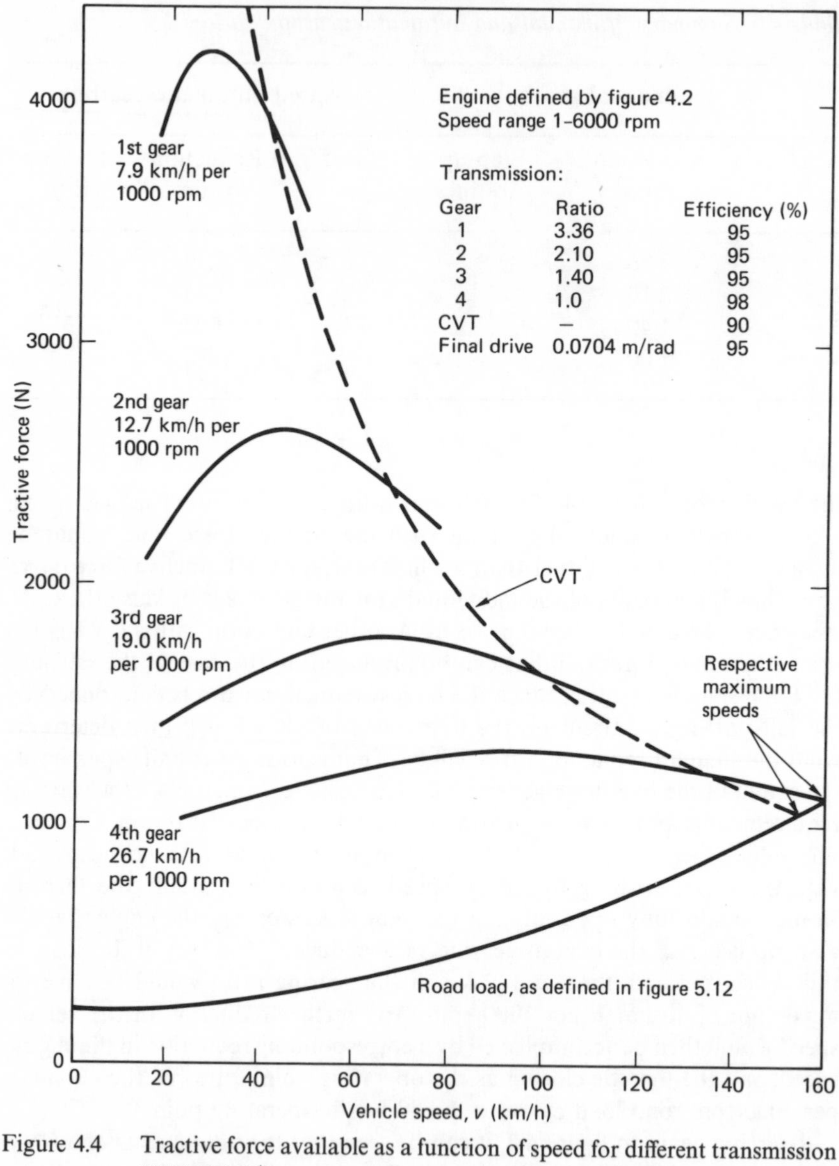

Figure 4.4 Tractive force available as a function of speed for different transmission arrangements

This discussion has not included automatic gearboxes since the torque converter characteristics are difficult to define. Furthermore, since the trend is now towards automatic gearboxes with 'lock-up' torque converters (as discussed in section 4.4), the steady-state treatment is the same as for manual gearboxes.

Table 4.6 The effect of a CVT on medium-speed fuel economy at 80 km/h

	CVT			Fixed ratios	
Span	5	6	> 6.6	Top (3.36)	Overdrive (4.2)
Power ($\dot{W}_b$) (kW)	11.67	11.67	11.67	10.72	11.06
sfc (kg/kWh)	0.42	0.375	0.36	0.515	0.475
Fuel economy (km/kg)	16.3	18.3	19.0	14.6	15.2

4.2.5 Matching to diesel engine characteristics

The discussion here has been developed around the fuel consumption map of a spark ignition engine. Diesel engines have significantly different characteristics, and these are most apparent in the larger engines used for commercial vehicles. In such engines the speed range may be restricted from 1000 to 2500 rpm, and this necessitates multi-speed (say 12 ratio) gearboxes.

The differences are not so great for car-sized diesels, but the differences are still worth discussion. The fuel consumption map for a light automotive diesel engine is shown in figure 4.5, and can be compared with figure 4.3. The principle differences are

(a) The reduced speed range–even at 4500 rpm the brake mean effective pressure (bmep) is falling rapidly.
(b) The peak bmep occurs at a lower speed, and is some 2 bar lower.
(c) The efficiency of the diesel engine is higher, with a smaller reduction in efficiency at part load operation.

The reduced speed range and peak bmep of the diesel, compared with a spark ignition engine, are due to the nature of the combustion in a diesel engine, and this was discussed in section 3.2.2

For the diesel engine defined by figure 4.5, with the road load curve shown in figure 5.12, the calculations for the spark ignition (SI) engine can be repeated. The results will just be summarised here:

(a) Gearing ratio to give 160 km/h at the maximum output speed (4500 rpm): 35.56 km/h per 1000 rpm.
(b) Swept volume necessary to achieve 160 km/h: 2090 cm^3 compared with 1300 cm^3 for the SI engine.

The difference between the fuel economy of the SI engine and the diesel

Table 4.7 Comparison of fuel consumption for a diesel engine and an SI engine with different transmission systems

	Steady-state fuel economy (km/kg)			
	Diesel	SI direct drive	SI overdrive	SI CVT
90 km/h	16.8	13.7	14.7	18.6
120 km/h	16.3	11.0	13.7	14.9

engine at 120 km/h is usually less than that shown in table 4.7. In the example developed here, the part load fuel consumption of the SI engine is particularly high. The diesel may also appear to offer a greater torque reserve than an SI engine, but this is not necessarily the case. At 120 km/h the torque reserve for the diesel is 50 N m (compared with 41.7 N m), but the higher gearing for the diesel of 35.56 km/h per 1000 rpm (compared with 26.67 km/h) means that the tractive force available at the driving wheels will be less. It can also be seen from figure 4.5 that there is less to be gained in using an overdrive ratio with diesel engines. Further comparisons will be made between spark ignition and diesel engines in section 4.6, where the matching and optimisation of powertrains is discussed.

4.3 Manual gearboxes

For the reasons developed in the previous section, there is an increasing trend towards using gearboxes with an 'overdrive' ratio. Initially such a ratio was provided by a separate overdrive unit, but the current levels of production justify five-speed gearboxes.

The gearbox illustrated in figure 4.6 is of the constant mesh type, with synchromesh on all forward gears. In a constant mesh gearbox, all the gears are permanently meshed, and the gears on the mainshaft can rotate freely.

It is not immediately apparent in figure 4.6 that the main shaft and input shaft are not directly connected, but can have relative rotation through a bearing. The direct drive is attained when Selector A moves to the left and couples the input and main shafts.

In this gearbox the layshaft is fixed and the gears rotate as a single unit about the layshaft. This 'dead' arrangement is less stiff than a 'live' layshaft in which the shaft and gears all rotate, but this arrangement is easier to assemble. To maintain the layshaft rigidity, the overdrive gear is overhung from the bearing, to shorten the distance between the bearings; this is acceptable since the fifth gear transmits smaller loads.

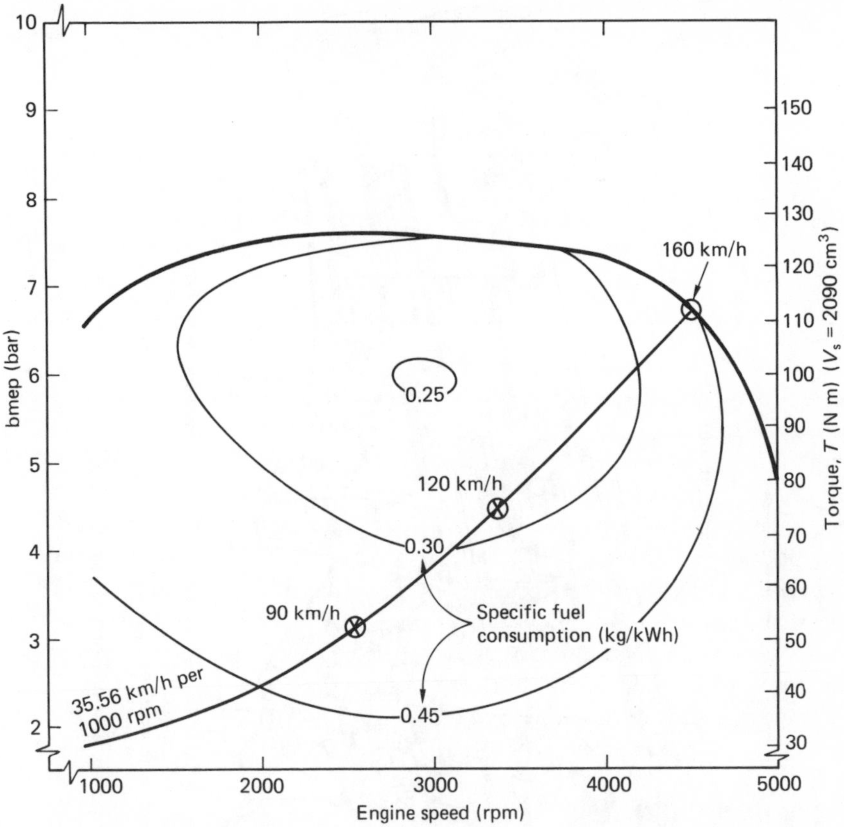

Figure 4.5 Diesel engine map with road load curve (defined in figure 5.12)

The drive through an intermediate ratio is obtained by a selector sliding a dog-clutch (that is, splined to the mainshaft) to engage with the appropriate gear on the mainshaft. Synchromesh devices are incorporated on to the dog-clutches, and baulk rings prevent the drive being engaged if the gear and dog-clutch are not running at the same speed. Synchromesh is much easier to achieve with constant mesh gearboxes than with sliding mesh gearboxes, in which gears are moved in and out of mesh.

The overdrive or fifth gear is selected in the same manner as the intermediate gears. Reverse rotation is obtained by an idler gear on a separate shaft; the reverse gear selector will only be able to slide this gear into mesh when the mainshaft is stationary.

For the sake of completeness, a brief description is given here of a separate overdrive unit. The overdrive is an additional gearbox mounted at the output of the main gearbox. The overdrive unit contains a set of epicyclic gears that can be engaged to speed up the output, with ratios typically in the range

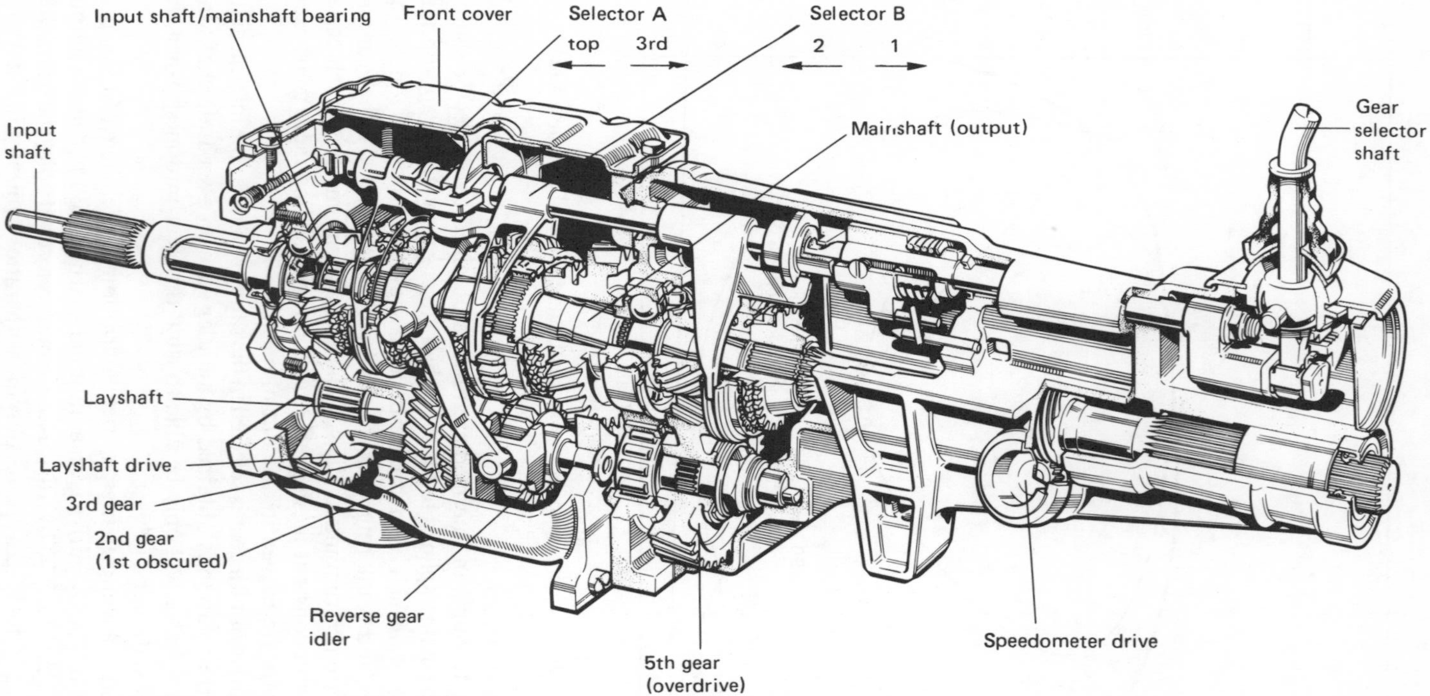

Figure 4.6 A Ford 5-speed gearbox. [Reproduced by permission of Ford Motor Co. Ltd]

0.7–0.8:1. The overdrive is engaged hydraulically through the operation of a solenoid valve; its use is limited to third and fourth gears in order to restrict the torque handling requirements of the overdrive unit.

An overdrive unit is shown in figure 4.7, and the main feature is the epicyclic (or planetary) gear unit with a cone clutch connected to the sun wheel. In direct drive, the inner surface of the cone clutch engages with the annulus, and the whole geartrain rotates. The drive is taken by the one-way clutch and cone clutch; when the load is reversed for engine braking, the load is taken solely by the cone clutch.

When the overdrive is engaged, the cone clutch engages with the outer casing to hold the sun wheel stationary. The drive input is to the planet wheel carrier, and this causes the annulus to rotate faster than the input shaft. The output shaft is permanently connected to the annulus, and can rotate faster than the input shaft by virtue of the one-way clutch. The overdrive ratio cannot be engaged in reverse since the one-way clutch will not allow the output shaft to rotate faster than the input shaft in the reverse direction. The one-way clutch may appear superfluous, but it permits the cone clutch to be sized for the torque associated with engine braking and overdrive as opposed to the much greater torques that can occur with the lower gear ratios.

4.4 Automatic gearboxes

The conventional arrangement for automatic gearboxes has been to use a

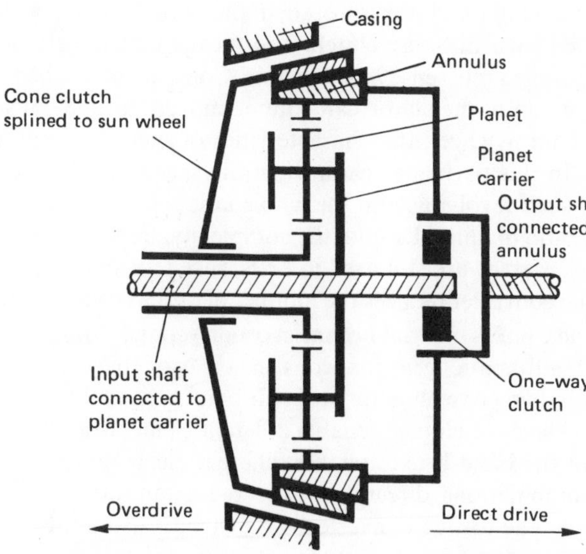

Figure 4.7 Laycock-de-Normanville epicyclic overdrive unit (shown in direct drive position), adapted from Abbott (1972)

three-speed epicyclic gearbox, with some form of fluid coupling to provide the equivalent of a clutch. The simplest form of fluid coupling is a fluid flywheel. The fluid flywheel in figure 4.8a has two vaned rotors that are in close proximity; one rotor is attached to the engine and the other is attached to the gearbox. At low speeds (idling) there is no motion transferred between the rotors, but as the speed rises above about 1000 rpm the relative slip falls to a few per cent. Since the only components subject to a torque are the rotors, the equilibrium of moments (or torques) dictates that the input and output torques are equal. However, a slip of 5 per cent (which is typical of normal operation) represents a power loss of 5 per cent, and to a first order this also represents a 5 per cent fuel economy penalty. The torque/speed characteristics for a fluid flywheel are shown in figure 4.8b.

Torque converters are somewhat more complex, since in addition to two rotors there is a stator (figure 4.8c). Since the stator is fixed it can provide a torque reaction, and the torque on the two rotors no longer has to be equal. The torque converter is equivalent to a radial flow pump and turbine in series, sharing the same stator. Like all such dynamic flow machines, the efficiency away from the design point is low. Torque converters can provide an appreciable torque multiplication, but the peak efficiency for the conditions shown is only 80 per cent (figure 4.8d). Obviously, even a 20 per cent power loss is unacceptable under steady-state conditions for two reasons: firstly, the associated increase in fuel consumption and, secondly, the need to dissipate the heat that is generated by the power loss.

This problem can be overcome by some form of locking between the two rotors, or by allowing the stator to rotate in the same direction as the engine. This is achieved by a one-way clutch, and means that the stator can only exert a torque of the same sense as the engine. Consequently, when the torque ratio falls below unity the stator can rotate, and the torque converter then behaves as a fluid flywheel; the characteristics of such a system are shown in figure 4.8e. In this particular case, when the speed rises above 1350 rpm the stator will start revolving, and the whole unit behaves as a fluid flywheel with a torque ratio of unity. Despite the complexity needed to make a torque converter efficient, they are still used in preference to a simple fluid flywheel, since the torque converter reduces the number of ratios needed in the gearbox.

Automatic gearboxes invariably consist of epicyclic geartrains; for example, a three-speed automatic gearbox consists of two epicyclic geartrains, a one-way clutch, two cone clutches and two band brakes that are controlled hydraulically. The cone clutches enable different elements in the geartrain to be driven, and the band brakes enable different elements to be fixed. With appropriate combinations, direct drive, two reduction ratios and reverse can all be obtained. The torque converter and gearbox also need to transmit a torque in the opposite sense to provide engine braking. However, a car with an automatic transmission cannot usually be tow started, since the hydraulic pump that provides oil for the actuation and control system is driven from

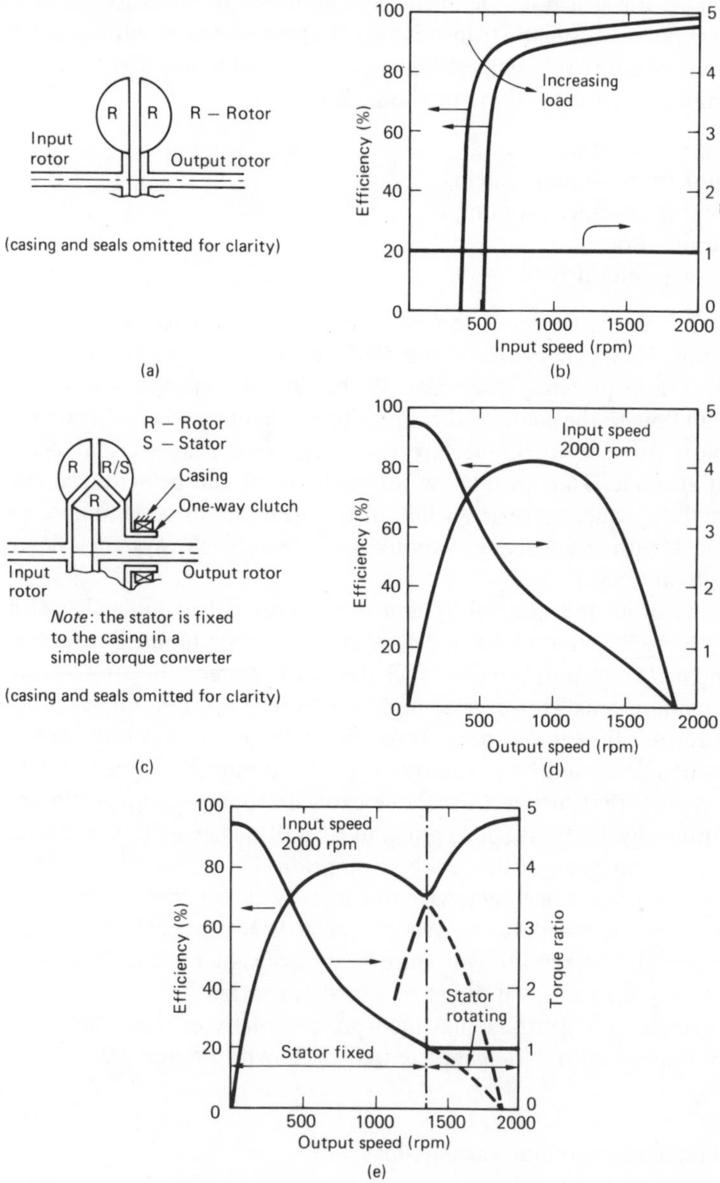

Figure 4.8 Types of fluid coupling and their characteristics: (a) kinematic diagram of a fluid flywheel, (b) fluid flywheel torque/speed characteristics, (c) kinematic diagram of a torque converter, (d) torque converter torque/speed characteristics, (e) the torque speed characteristics for a combination of a torque converter and a fluid flywheel

the input to the gearbox. Detailed explanations for the working of many different types of automatic transmission can be found in Newton *et al* (1983).

A variety of approaches have been used in the past to improve the efficiency of automatic gearboxes; these methods include:

(a) torque converter locking
(b) shunt transmission systems
(c) variable capacity oil pumps
(d) overdrive ratios
(e) free-wheeling drives.

If the torque converter is locked, the losses associated with slip under steady-state operation can be eliminated; this gives an improvement in fuel consumption of perhaps 5 per cent. With a shunt transmission system, the power can flow to the gearbox through either the torque converter or a direct mechanical drive. Both inputs are to an epicyclic gear set that acts in a differential mode. Thus in the low ratios the drive can be arranged entirely through the torque converter, while in top gear the drive is almost purely mechanical (with an efficiency of about 97 per cent); such a system is described by Dorgham (1982).

The actuation and control system are operated hydraulically, with the power coming from an oil pump driven at the gearbox input. Conventionally the pump has a constant capacity, such that the required flow can be produced at low engine speeds. However, at high engine speeds excess flow is produced and the surplus flow is dissipated through a relief valve; this source of power loss can be eliminated by a variable-capacity pump. Koivunen and LeBar (1979) report that using a variable-capacity pump reduces the power consumption by 1 kW at 3000 rpm, and overall this leads to something like a 5 per cent reduction in the fuel consumption.

The use of a free-wheeling device or one-way clutch enables the engine to return to its idling speed during over-run. The disadvantage is that the engine braking effect is lost, but this will cause least inconvenience if the free-wheeling device is only operative on direct or overdrive ratios.

An example of a gearbox that incorporates many of these devices is the Ford four-speed automatic gearbox that is shown in figure 4.9.

4.5 Continuously variable transmissions

4.5.1 Introduction

In order for a continuously variable transmission (CVT) to produce an improvement in fuel economy, it needs to be efficient and have a wide span (range of ratios). There are two significant types of CVT–the Van Doorne belt system and the Perbury traction system–both of which have been the

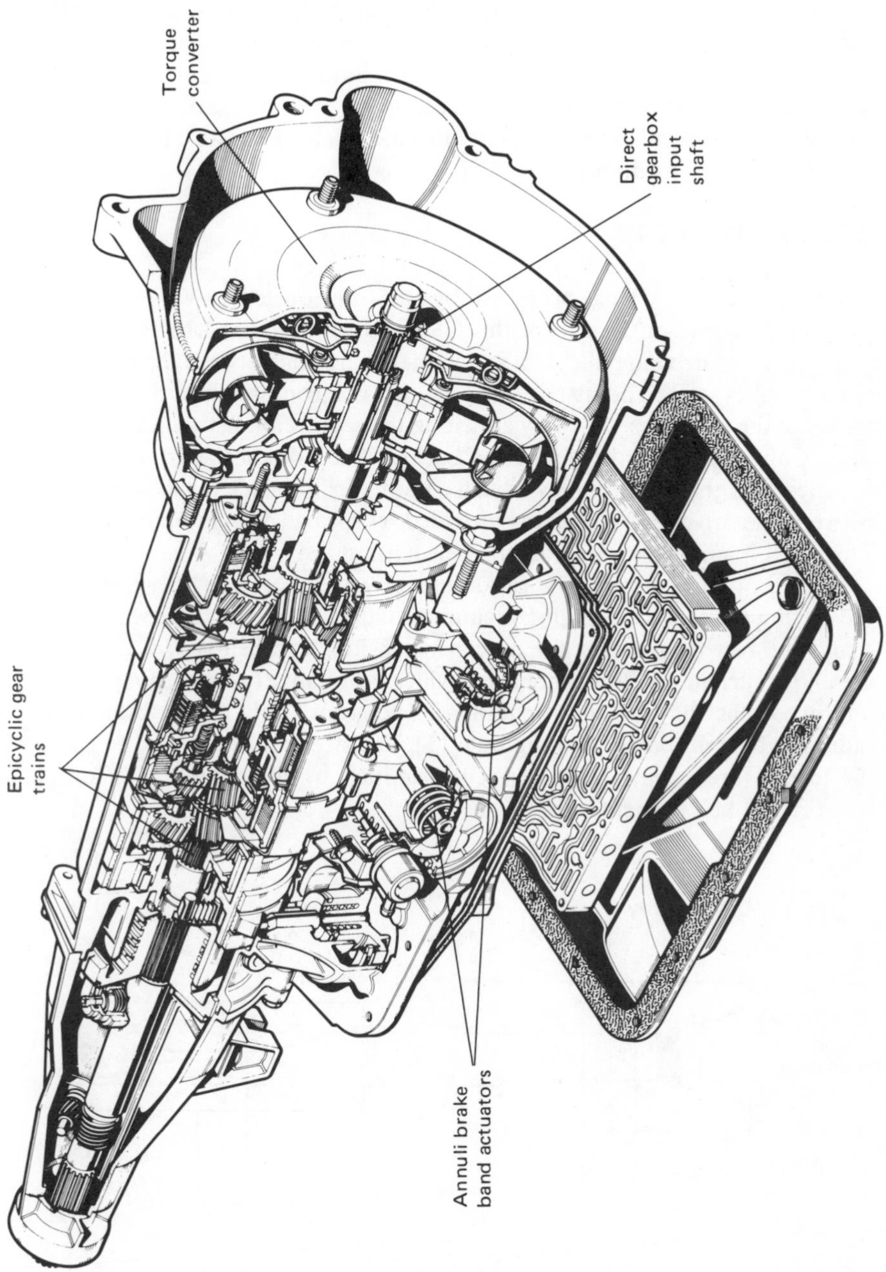

Figure 4.9 Ford A4LD 4-speed Automatic Transmission. [Reproduced by permission of Ford Motor Co. Ltd]

subject of much development work. The belt system is most suited to low power applications, and has particular advantages for front wheel drive vehicles. The Perbury system has been built in much larger sizes, and lends itself to conventional in-line engine gearbox installations; prototypes have been installed for testing in delivery trucks. In both cases the control strategy is of great importance if maximum fuel economy, maximum acceleration and engine braking are all to be obtained. Indeed, a microprocessor-based control system is almost inevitable.

4.5.2 Van Doorne CVT

The Van Doorne CVT has been used by Daf (now owned by Volvo) on cars since 1955. The essential part of this CVT, shown in figure 4.10, is a pair of conically faced driving pulleys, in which the separation between the two sides of each pulley can be adjusted. Since the belt is incompressible, the effective radius of the pulleys is varied by adjusting the separation between the pulley sides. Conventionally, the separation between the sides of the driving pulley is varied, and the driven pulley adjusts automatically, since the axial loading between the driven pulley sides is provided by springs or hydraulic pressure. This axial loading also controls the belt tension.

On the original system, the ratio control was by centrifugal weights on the driving pulley and an engine vacuum actuator. The ratio range was about 4 and the belt transmitted the power in tension. The belt-type CVTs that are under current development have a wider ratio range (approaching 6) and improved efficiency (in the range 86–90 per cent); this is the result of a radically different belt design (figure 4.11). The steel bands are used to carry the tensile forces, but the power is transmitted by the compressive forces between the belt elements.

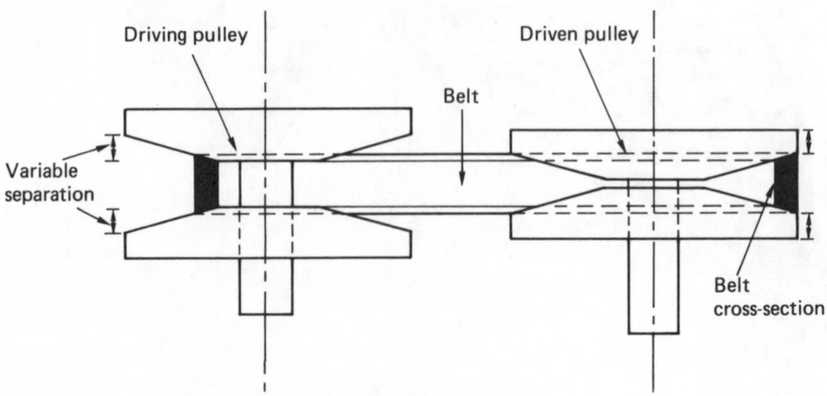

Figure 4.10 Principle of the continuously variable belt transmission, in the maximum reduction position

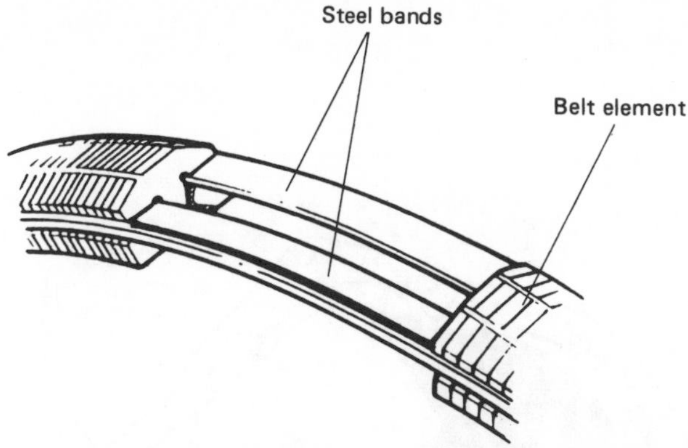

Figure 4.11 CVT belt construction, Newton *et al.* (1983). [By permission of the publishers, Butterworth & Co. (Publishers) Ltd]

A complete Ford CTX (continuously variable transaxle) is shown in figure 4.12, and a description of some of the development work is given by Hahne (1984). The belt drive is particularly suited to compact front wheel drive vehicles, since it provides a convenient means of connecting the engine and differential. For the CTX shown in figure 4.12, the input drives the planet carrier and the inner part of the forward drive clutch. The outer part of the forward clutch is connected directly to the driving pulley, and this provides the path for power transmission. The sun gear is also connected directly to the driving pulley, and if the reverse clutch brakes the annulus, then the planet carrier will drive the sun gear in reverse. These clutches are of the wet type with hydraulic actuation, and have a more predictable performance during their life than dry friction clutches. A torque converter could have been incorporated, but the associated losses would remove the benefits of a CVT. The Ford CTX is controlled hydraulically and responds to five inputs: shift lever position, accelerator pedal position, pulley ratio, engine speed and primary pulley speed. The transmission is designed for a torque of 125 N m.

4.5.3 Perbury CVT

A key part of the Perbury CVT is the variator, a tilting roller assembly that is shown in figure 4.13; this had already been used on cars in the 1930s by Hayes. The variator consists of three discs, with the outer pair connected. The discs have a part toroidal surface on their inner faces, upon which the spherically faced rollers roll. The rollers can rotate about their own axes and the inclination of these axes is varied, but the carrier for the rollers is fixed. The inclination of these rollers is varied by a control sleeve acting through rockers, and this leads to the continuously variable ratio changes in the variator.

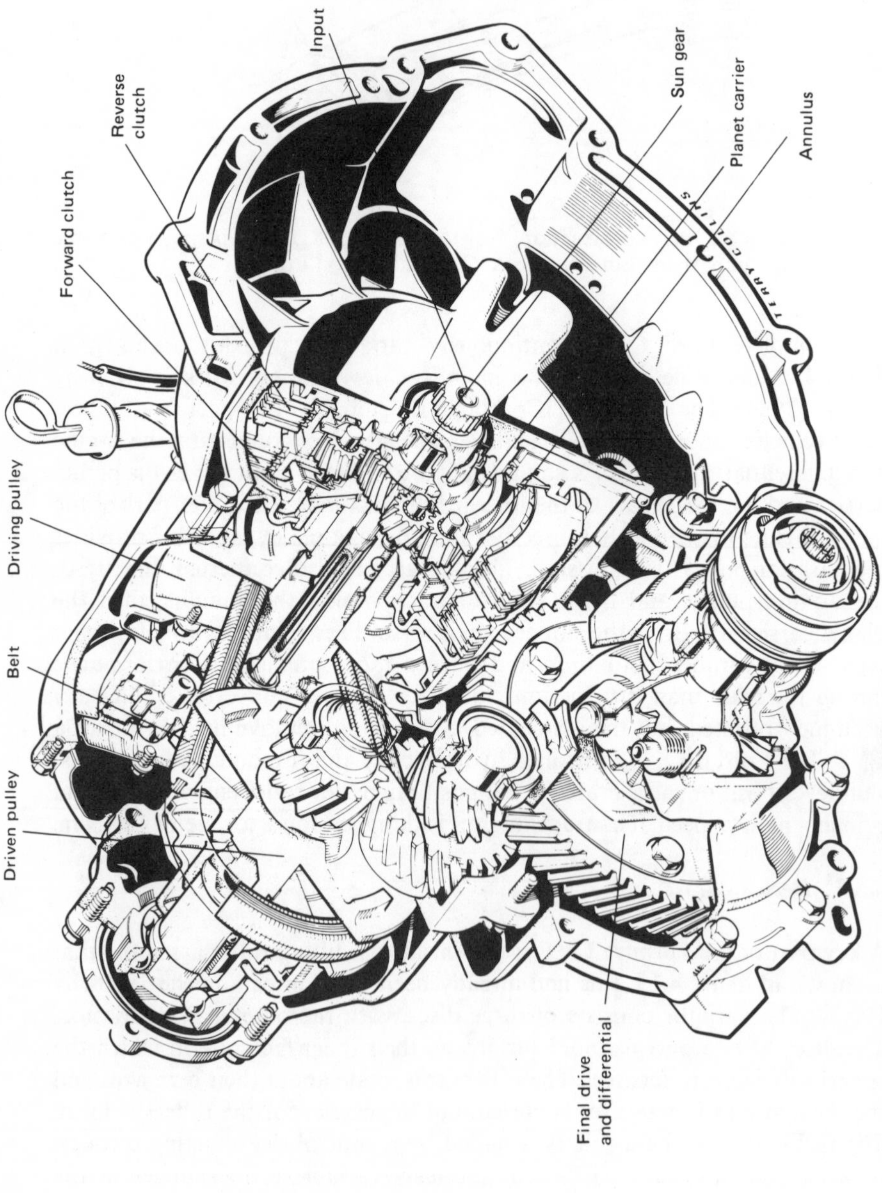

Figure 4.12 Ford continuously variable transaxle. [Reproduced by permission of Ford Motor Co. Ltd]

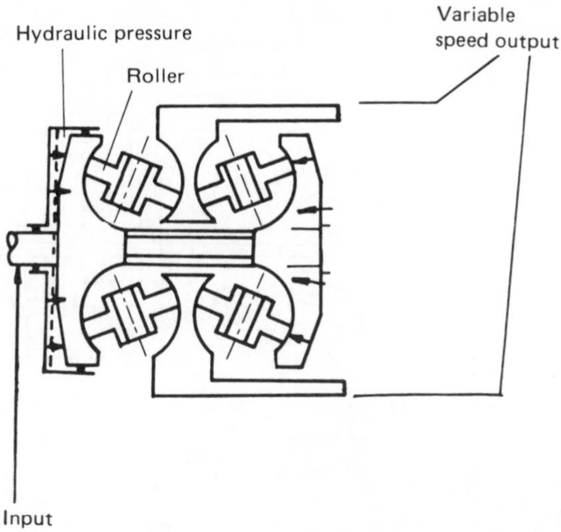

Figure 4.13 Variator assembly, Greenwood (1984). [Reprinted by permission of the Council of the Institution of Mechanical Engineers]

To prevent wear, lubrication is needed and an elasto-hydrodynamic oil film exists between the rollers and discs. Relative slip (typically 1–2 per cent) is inevitable, since it is this that produces the shear in the oil film, which transmits the tractive forces that are tangential to the variator surfaces (hence the term 'traction drive'). Even with specially formulated oils the coefficient of friction (or in this case traction) is very low, thus very high contact forces are needed between the discs and rollers. This is achieved in a system with internally balanced forces, by applying hydraulic pressure to the outer face of the driving disc that is splined to the input shaft. Since the pressure can be varied, the unit life is improved by reducing the contact forces when the variator is lightly loaded.

The control of the rollers is critical if the three rollers on each side are to transmit equal loads and operate at the same ratio. This is achieved automatically in the Perbury drive, by ingenious design of the roller inclination actuation mechanism. Three disadvantages that the basic variator shares with the Van Doorne belt system are that: the ratio range is less than about 5, there is no ratio that will give zero output (to eliminated the need for a clutch), and there is no reverse ratio. All these shortcomings are overcome in the Perbury transmission, by using a shunt transmission system with an epicyclic gearbox.

The key elements of the Perbury transmission are shown in figure 4.14. There are two clutches, one of which is engaged for the high-speed regime and the other that is engaged for the low-speed regime. The various modes will now be considered assuming a constant input speed.

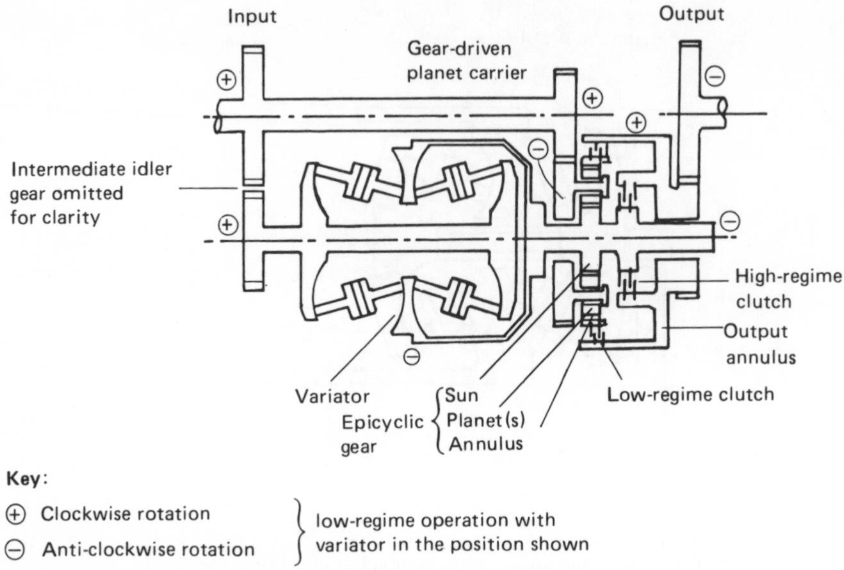

Figure 4.14 Essential elements of the Perbury transmission, Greenwood (1984). [Reprinted by permission of the Council of the Institution of Mechanical Engineers]

(a) *High regime.* When the high-regime clutch is engaged, the epicyclic geartrain is not used and the variator output is coupled directly to the output shaft via the high-regime clutch, the output annulus and the output gear. With the rollers in the position shown, the output speed is at its greatest; as the roller inclination is reduced, the output speed will also reduce.

(b) *Regime changeover.* The gear ratios have been chosen so that when the output from the variator is at its lowest speed (the opposite to that illustrated) the gear-driven planet cage and the sun gear are travelling at the same speed. Thus the epicyclic annulus and the output annulus are also travelling at the same speed. This is called the synchronous ratio, and it enables a smooth transition between the regimes to occur by engaging the low-regime clutch and disengaging the high-regime clutch or vice versa.

(c) *Low regime.* The epicyclic geartrain can act as a differential. If the input speed to the gearbox is constant, then the gear-driven planet carrier will rotate the planets about the sun gear at constant speed. If the output speed of the variator is now increased, the differential action of the epicyclic will reduce the speed of the epicyclic annulus and the gearbox output. As the variator output is increased yet further, the gearbox output is reduced and a point will be reached at which the gearbox output speed is zero; this is termed 'geared neutral'. If the variator output is increased further, then the gearbox output will reverse in direction; the maximum reverse speed will occur in the variator position shown in figure 4.14.

The overall Perbury transmission characteristics are shown in figure 4.15. In the high regime the drive is occurring solely through the variator, while in the low regime the drive occurs through the variator and a direct connection into the epicyclic geartrain. The low regime is thus a shunt transmission system, with the epicyclic gearing acting as a differential, in order to combine the two inputs. Since the geared neutral provides zero speed output or infinite

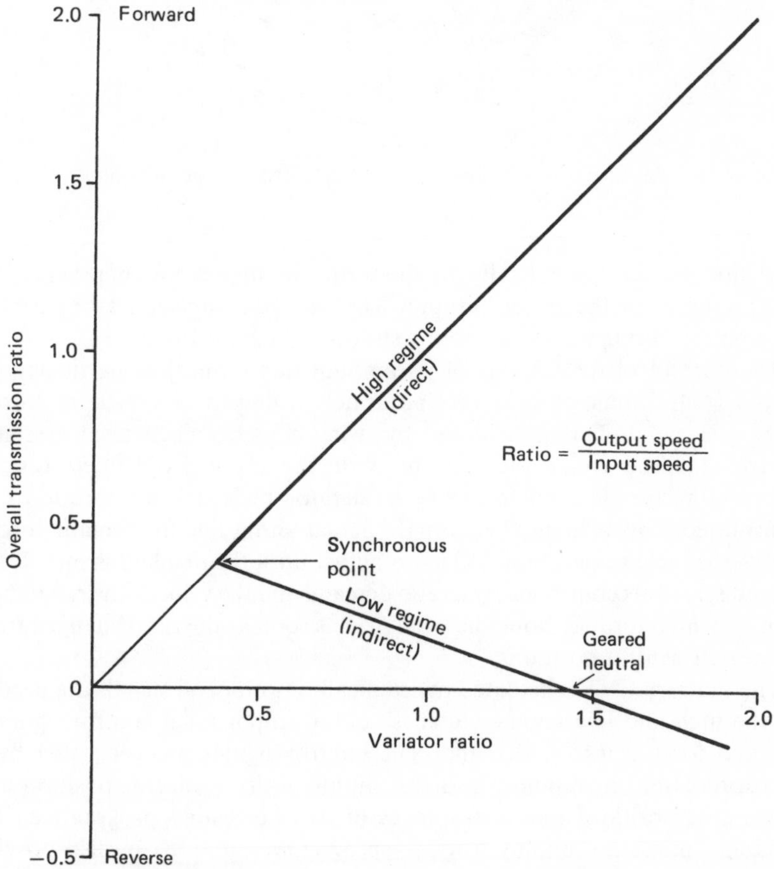

Figure 4.15 Operating regimes of the Perbury transmission, Ironside and Stubbs (1981). [Reprinted by permission of the Council of the Institution of Mechanical Engineers]

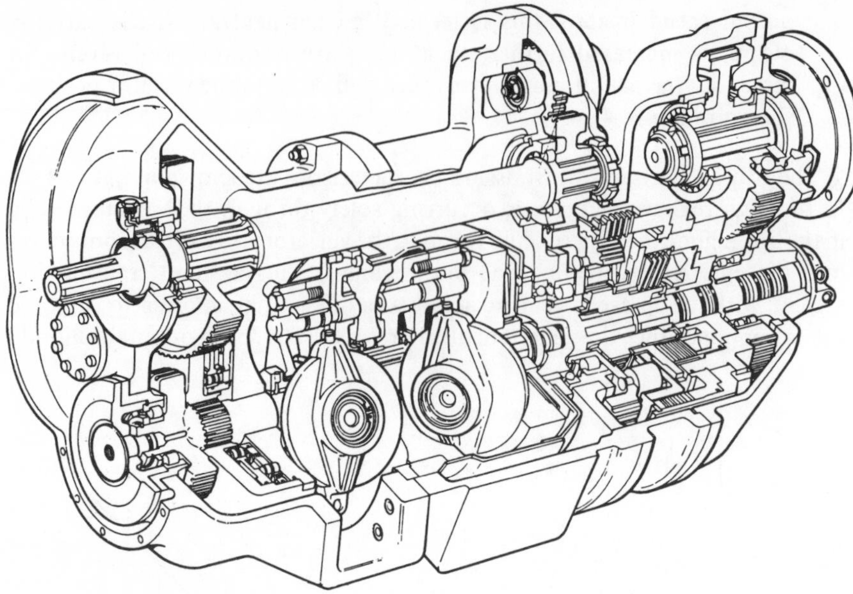

Figure 4.16 Perbury transmission system. [Reprinted with permission from *Design Engineering*]

reduction, there is theoretically infinite torque multiplication; in practice, the only limitation on the torque output will be the torque capacity of the variator. A complete Perbury transmission is shown in figure 4.16.

The control of the CVT is of paramount importance; in particular the geared neutral must be held very accurately if unwanted vehicle motion is to be prevented. This is achieved by using a hydro-mechanical feedback control for the roller inclination, with the input hydraulic pressure representing the demand from the accelerator pedal for the torque to be transmitted. Any difference between the actual torque and the demand torque causes the inclination of the rollers to adjust until the demand is met. If the transmission was controlled by a ratio demand and the vehicle was restrained, then very high torques would be produced as soon as the ratio changed from the geared neutral position.

The Perbury CVT is actuated hydraulically: two on/off signals are needed for the high and low regime clutches, and a proportional control signal is required for the torque demand. The control signals are generated by a microprocessor, in response to driver inputs of drive selector position and accelerator position. The microprocessor can be readily programmed for different optimum economy torque/ speed curves, and it can also provide additional facilities; for example, the torque can be limited at low vehicle speeds to prevent wheel spin.

Additional controls and sensors are also needed. During deceleration the

CVT must not change ratio so rapidly that the engine decelerates at the expense of accelerating the vehicle. Also during braking it is necessary to sense if the driven wheels become locked.

The Perbury transmission described here has been developed by Leyland Vehicles for use in buses and delivery trucks; further details can be found in Ironside and Stubbs (1981) and Greenwood (1984). The Leyland unit is rated at 375 kW, and the fuel economy gains are expected to be about 10 per cent, in urban traffic.

A further use of the Perbury CVT is in regenerative braking. Energy that would otherwise be dissipated by braking is stored as kinetic energy in a flywheel. A flywheel developed by BP is described by Beevers (1985); the flywheel has a mass of 160 kg and can revolve at speeds of up to 16 000 rpm, at which it stores 1.4 MJ (equivalent to 0.04 litres of diesel fuel). Since the flywheel needs to be decelerated independently of vehicle speed, a CVT is the requisite solution. The flywheel deceleration is such as to be able to deliver 160 kW. A comparable system described by Garrett (1986) uses a hydraulic pump/motor with an accumulator that can store 0.8 MJ at pressures up to 350 bar. Fuel savings of 30 per cent are claimed for city bus routes.

4.6 Powertrain optimisation

Powertrain optimisation is evidently not possible by analytical solution of closed form equations; indeed, the simple examples developed in section 4.2 are useful as illustrations of a method, but a method that would quickly become tedious. Thus powertrain optimisation lends itself to computer modelling. The requirements for a model would include the following.

For steady-state operation

(a) Access to engine maps showing the torque output and specific fuel consumption, and a method of interpolation.
(b) A method for deducing wind resistance as a function of area, drag coefficient, ambient conditions and vehicle speed.
(c) A method for deducing rolling resistance as a function of vehicle weight, speed and tyre type.
(d) A way of specifying the gearing ratios and their efficiency as a function of the operating condition.

For transient operation

(e) Moment of inertia of the engine and moment of inertia of the gearbox in different gears.

(f) Time constants for gear changing and engine response (especially if turbocharged).
(g) Modifications to the engine map while the load/speed operating point is changing.

For cycle simulation

(h) A method of specifying the required velocity as a function of time.
(i) Algorithms for deducing the gear changes for different driving characteristics (for example, economy or sport).
(j) Allowance for the fuel consumption on over-run.

Emissions

(k) The steady-state engine emissions.
(l) Allowances for the effects of transient operation during acceleration and braking.

Such a computer model would have a means of specifying the basic vehicle to be 'tested' and the changes to be made during the optimisation. Examples of this approach can be found in many papers, such as Porter (1979), Thring (1981) and Lorenz and Peterreins (1984). This approach enables trade-offs to be investigated without recourse to the testing of numerous vehicles. Obviously it is still necessary to test the final vehicles in order to validate the model.

An example that might be investigated is the trade-off between fuel consumption and performance for a given vehicle, with different capacity engines and final drive ratios. The results will depend on the definition of performance and fuel consumption. The performance may be specified as the time taken to travel 400 m from rest, or to accelerate between specified speeds. Equally, the fuel consumption may be at a specified speed or speeds, over a specified cycle, or some weighted combination.

Figure 4.17 shows the predicted trade-off between fuel economy and performance for a General Motors 'X' car with different transmission ratios and engine sizes. The fuel economy is a 55/45 weighted average of the EPA (American Environmental Protection Agency) urban and highway driving cycles.

For any particular fuel economy requirement there is an optimum engine size and transmission ratio. With reference to points A and B on figure 4.17, both represent different engine/transmission combinations with the same fuel economy. However, it is the vehicle with the larger engine that offers the better performance.

The optimisation will not depend solely on the final drive ratio and engine size, but also on the intermediate gear ratios. As a simple example, consider

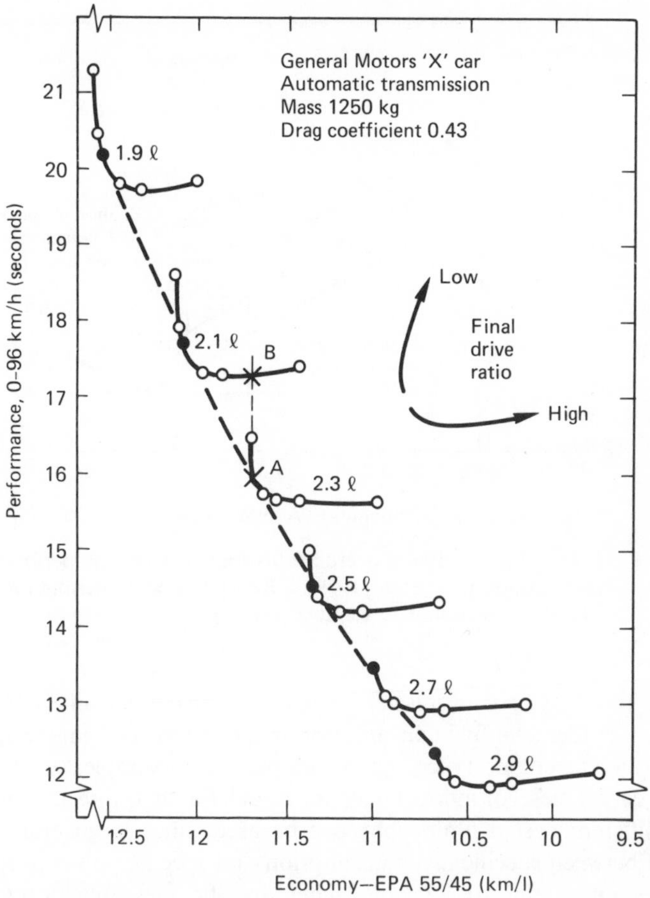

Figure 4.17 The effect of engine displacement (litres), and final drive ratio on performance and fuel economy, adapted from Porter (1979)

the steady-speed fuel economy and acceleration from 0 to 100 km/h. The steady-state fuel economy will be good if there is a low reduction ratio final drive. This will not aid acceleration, but can be compensated for by having high reduction ratios in the intermediate gears. In other words, a gearbox with a large span is needed; this is illustrated more generally by figure 4.18. However, as the ratio span increases, the number of discrete gear ratios needed increases, if progression through the gears is to remain smooth.

The number of gear ratios that can be justified will also depend on economics. In the case of manual transmissions there will be a finite number of ratios that a driver can be expected to manage, for two reasons:

(a) Fatigue associated with continual gear changes.
(b) Knowing which ratio to be in for a particular road condition.

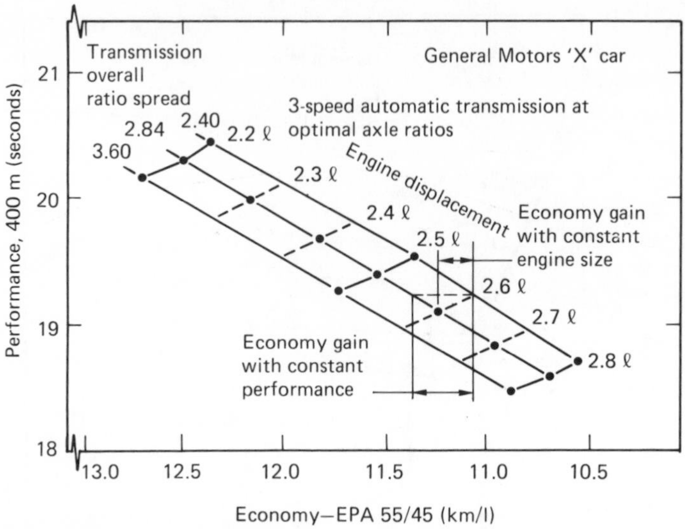

Figure 4.18 Effect of transmission overall ratio spread (span) on performance and fuel economy, Porter (1979). [Reprinted with permission © 1979 Society of Automotive Engineers, Inc.]

This discussion of transmission ratios has considered different capacity engines, with identical fuel consumption maps in terms of engine speed and bmep. Different engine types can be adopted–for example, spark ignition, direct injection diesel or indirect injection diesel. Furthermore, for each engine type a different design philosophy can be used, since in general there is a trade-off between specific fuel consumption and specific power output (high output engines tend to have a higher specific fuel consumption). The optimisation now needs to include taxation policy, since different countries have their own taxation policies for fuel and engine size. If a country taxes vehicles according to engine displacement, then manufacturers have an incentive to produce high output engines. While this makes the engines lighter for a given power, the fuel savings associated with vehicle lightness are unlikely to offset the lower efficiency of such engines.

The characteristics and fuel economy of spark ignition engines and diesel engines have been discussed respectively in chapters 2 and 3, but for completeness the main conclusions are summarised here:

(a) Engine friction and parasitic losses (for example, the water pump) are virtually independent of load. Thus at part load operation, engine friction is very significant, and is the main reason for the reduced part load fuel economy of all engines.
(b) Engine friction rises very rapidly with speed, and the smaller speed range of diesel engines is one reason for their better fuel economy.

(c) At part load, the output of spark ignition engines is regulated by throttling the air flow; this irreversible process dissipates work.
(d) At part load, the corresponding cycle for diesel engines becomes more efficient, and this compensates for the increasing significance of engine friction, and the part load fuel economy of diesel engines is reasonable.
(e) Engines designed for high outputs at high speeds have valve timings with induction and exhaust systems that are notably less effective at lower speeds–the normal operating regime.

The way that the engine design philosophy affects vehicle fuel economy is discussed by Radermacher (1982). BMW compared engines with identical power ouputs but different swept volumes and torque curves; the engine with the larger swept volume and smaller speed range was termed the Eta engine. A comparison of the fuel economy maps is presented in figure 4.19. For the same vehicle, a different final drive ratio is used to give the same top speed; this leads to the two different tractive resistance curves. Comparison of the thermal efficiency contours (proportional to the inverse of the specific fuel consumption) shows that the maximum efficiency of the Eta engine is 32 per cent compared with 30 per cent for the standard engine. Analysis of the engine test results showed that the improved efficiency was attributable almost entirely to the increased mechanical efficiency of the Eta engine, a consequence of its slower running speeds.

The road load curves on figure 4.19 show the way in which the engine efficiency at typical operating points is improved; these are of much greater significance than the improvement in peak efficiency. It can also be seen that the torque reserve is still sufficient to provide good driveability. The improvement in fuel consumption by adopting the Eta engine is 12 per cent for a mixture of city, country and motorway driving.

Radermacher also describes subsequent development work to further reduce engine friction, along with a comparison between the Eta engine and diesel engines. Figure 4.20 shows the fuel consumption of the Eta engine and diesel engines, relative to a standard engine at different speeds. The fuel economy advantage associated with diesel engines is lost at higher speeds. If allowance is made for the higher energy density of diesel fuel, then the advantages of a diesel engined vehicle are even less significant. However, it must be remembered that fuel is bought and taxed on a volumetric basis.

4.7 Conclusions

In order to obtain a vehicle with good driveability, the transmission requirements have to be matched to the engine and vehicle characteristics. This matching is illustrated by an extended example in section 4.2, and the requirements for this to be done by a computer program are presented in section 4.6.

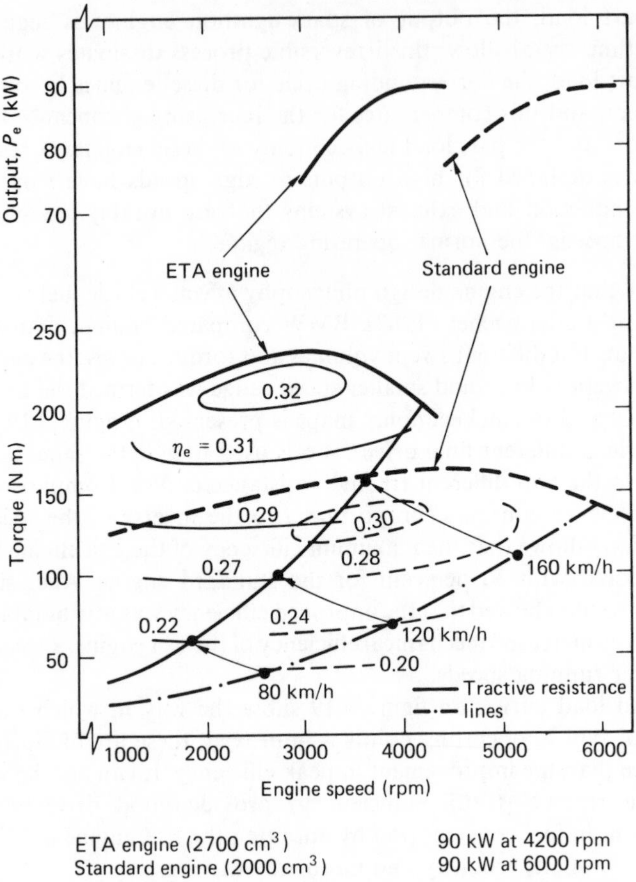

Figure 4.19 Comparison between the ETA engine and standard engine, showing the different tractive resistance lines and thermal efficiency contours. Radermacher (1982). [Reprinted by permission of the Council of the Institution of Mechanical Engineers]

The characteristics of manual gearboxes, automatic gearboxes and continuously variable transmissions were discussed respectively in sections 4.3, 4.4 and 4.5. In each case a wide span is needed for good fuel economy, and due account needs to be taken of the reduced efficiencies when any form of indirect gearing is used.

In the case of automatic transmission (with either discrete or continously variable ratios) the gear ratios are selected hydraulically and controlled hydro-mechanically. Such systems tend to be complex and ingenious, but even so they cannot necessarily provide optimal control. The use of microelectronics would appear obvious, with the ever-decreasing cost and

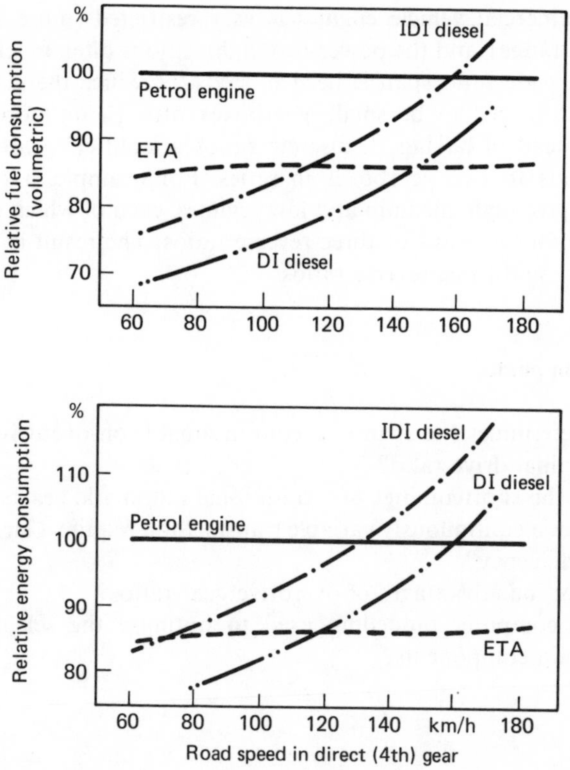

Figure 4.20 Comparison of the energy and fuel consumption for different engine types, Radermacher (1982). [Reprinted by permission of the Council of the Institution of Mechanical Engineers]

increasing use for engine control. It is possible to imagine two approaches to the control:

(a) to store in memory the optimum operating point for every conceivable load/speed condition, or
(b) to use a form of adaptive control that automatically seeks the optimum operating condition.

The second approach is obviously more complex, but would allow for inter-vehicle variation and the effects of wear. However, neither system has been adopted on a commercial basis, since the integrity and reliability of an electronically controlled system are more difficult to prove and maintain. In particular, the microelectronics would be susceptible to electro-magnetic interference, and the effects of a misjudged ratio change could be disastrous.

Commercial vehicle transmissions have not been dealt with explicitly; the principles in matching are, of course, the same. However, since the speed range of commercial vehicle engines is very restricted (more so than the vehicle speed range), and the power-to-weight ratio is often much inferior to a car, then a wide ratio span is needed, say 12:1. Since the steps between gear ratios still need to be small, gearboxes with 12 or more ratios are common. Instead of having 12 discrete ratios a 'splitter' gearbox is used, which amounts to two gearboxes in series. For example, a twin splitter gearbox can give high, medium and low regimes, each of which can be used in one of the four forward or three reverse ratios. The result is then twelve forward ratios and three reverse ratios.

4.8 Discussion points

(1) What determines the span of a conventional (non-overdrive) gearbox, and the final drive ratio?
(2) Identify the shortcomings of a traditional automatic gearbox.
(3) Why does a continuously variable transmission require a large span and a high efficiency?
(4) What are the advantages of overdrive gear ratios?
(5) How is computer modelling used to optimise the selection of the powertrain components?

5 Vehicle Aerodynamics

5.1 Introduction

Motor vehicle aerodynamics is a complex subject because of the interaction between the air flow and the ground, and the complicated geometrical shapes that are involved. Fuel economy is obviously improved by reducing the aerodynamic drag, and the benefits can be calculated easily for constant-speed operation. However, the actual benefit for normal use will be less, since drag reduction does not significantly reduce the energy required for acceleration at normal speeds.

Drag reduction is not the only aerodynamic consideration. The air flow will also affect the aerodynamic lift forces, and the position of the centre of pressure, both of which can have a profound effect on the vehicle handling and stability. While the presence of the ground has only a slight effect on the drag forces, it has a profound effect on the lift forces.

The aerodynamic designer also needs to consider the way in which the air flow controls the water and dirt deposition patterns on the glass and lamp surfaces. It is also important to minimise any wind noise, and to design for the ventilation flows. The air flow for engine cooling is the most significant, and the air flows for the passenger compartment, brake and transmission cooling are all much less significant.

The comparison of drag data from different tests should normally be avoided, since the absolute values will depend on the details of the experiment; the reasons for this are explained in the next section. However, it is valid and appropriate to examine the changes in drag as a result of changes to the vehicle shape in a given sequence of tests.

In general, vehicles are still designed by body stylists, and aerodynamicists then develop refinements to the shape in order to give reductions in drag and other aerodynamic improvements. However, since the stylists are becoming more conscious of the desirability of low drag vehicles, the basic vehicles shapes are also becoming more streamlined.

Because of the highly complex, three-dimensional, time-variant nature of the flow around a vehicle, it is not possible to computer-model the complete flow fully. Numerical techniques can be used to predict the main features of a flow or to examine some small aspect of the flow in a key area. This means that the experimental techniques applied to models in wind tunnels are very important, and these are discussed in the next section, after a treatment of the fundamentals of vehicle aerodynamics. Since experimental testing is time consuming and expensive, as much refinement as possible is achieved by computer modelling; this is increasing in importance with the ever-greater capabilities of computers and their programs.

Since passenger and commercial vehicles have such radically different shapes, they are the subjects of separate sections. Commercial vehicles are much less streamlined than passenger vehicles and, in general, aerodynamic drag is less significant since the speeds are lower and the rolling resistance is more significant because of the greater weights. An additional complication that is common with trucks is the tractor/trailer combination. The aerodynamic behaviour depends on the spacing between the two bodies. The two extremes are zero separation, where the behaviour is that of a single body, and infinite separation, where there is no 'slip-streaming' effect and the drag will be that of the two bodies in isolation.

5.2 Essential aerodynamics

5.2.1 Introduction, definitions and sources of drag

Consider a vehicle moving in a straight line on horizontal ground; the air flow is dependent on the vehicle speed and the ambient wind, as shown in figure 5.1. The wind has a non-uniform velocity profile because of the local topography and the earth boundary layer, and in general the velocity will be fluctuating in both magnitude and direction. The aerodynamic forces and moments act at the centre of pressure; the aerodynamic moments have been omitted from figure 5.1 for clarity. Unlike the centre of gravity, the centre of pressure is not fixed, but depends on the air flow; the centre of pressure tends to move forwards at higher velocities. The aerodynamic forces are resolved in the manner shown by figure 5.1, since it is the component in the direction of the vehicle motion that has to be overcome by the tractive effort, not the component in the direction of the air motion.

Also shown in figure 5.1 is the lateral force coefficient centre–the centre of action for the lateral force coefficients from the front and rear tyres. For stable operation at all speeds the lateral force coefficient centre has to be behind the centre of gravity [Ellis (1969)]. As with the centre of pressure, the centre of the lateral force coefficient is not fixed, but will depend on the load transfer characteristics of both axles and the effects of traction at the driven axle. The vehicle will be stable if the centre of pressure is behind the

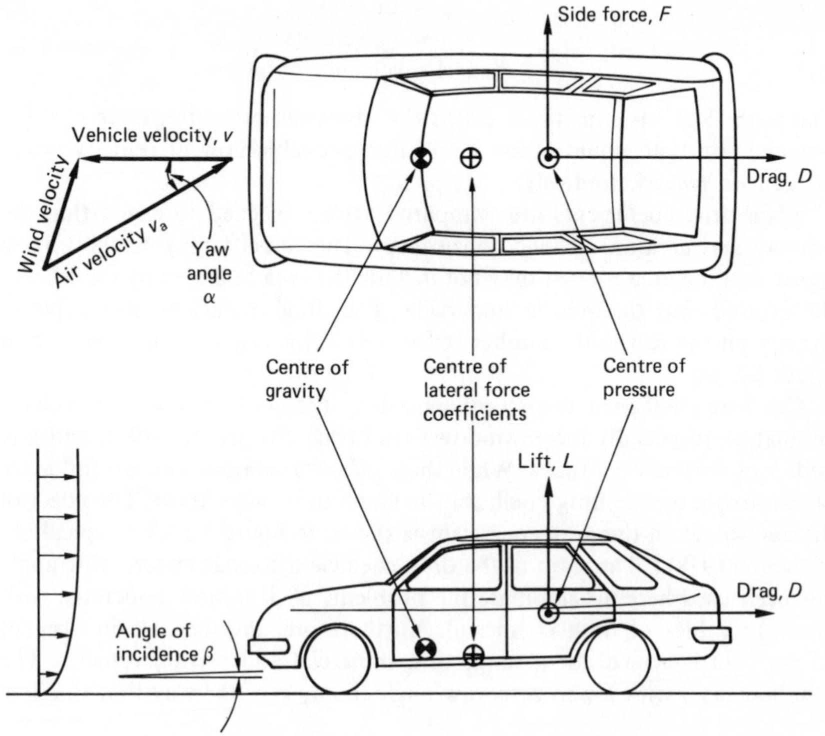

Figure 5.1 Aerodynamic forces on a vehicle in a real environment

lateral force coefficient centre. If the centre of pressure is in front of the centre of gravity, then a dynamic instability can arise: any divergence from the desired course introduces a turning moment about the centre of gravity, which tends to increase the divergence further. This can be alleviated by changing the slip angles of the tyres, and thus the position of the lateral force coefficient centre. The effect of aerodynamics on vehicle stability is described by Ward (1985) and discussed in great detail by Buchheim *et al.* (1985).

Since the vehicle and air velocity are not co-linear, there is a yaw angle, α, and a resultant side force. The lift force is a result of the asymmetrical flow above and below the vehicle, an effect that is evidently going to be influenced very strongly by the presence of the ground and the angle of incidence, β (defined in figure 5.1).

The drag and lift characteristics of a body are described by the dimensionless drag and lift coefficients, C_d and C_l. These are defined by the following equations:

$$\begin{aligned} \text{Drag, } D &= \tfrac{1}{2}\rho v^2 A C_d \\ \text{Lift, } L &= \tfrac{1}{2}\rho v^2 A C_l \end{aligned} \tag{5.1}$$

where ρ = air density
v = air velocity
and A = vehicle frontal area.

The term '$\frac{1}{2}\rho v^2$' is sometimes called the 'dynamic pressure', since it is the pressure rise that would occur if an incompressible flow of velocity v was brought to rest frictionlessly.

When drag coefficients are compared, care is needed to check that the velocity and area are defined consistently. The velocity may be the vehicle speed, and the area may or may not include the area bounded by the wheels, the ground and the vehicle underside. The drag coefficient also depends slightly on the Reynolds number (effectively velocity), and this is shown in figure 5.2.

The drag coefficient is also influenced by the cooling flows, the vehicle ventilation (especially if the windows are open), the ground effect, and any additions such as roof racks. While these effects are important, the influence of yaw angle on the drag coefficient is much more significant. The effect of the yaw angle on the drag coefficient is shown in figure 5.3 for a typical car [Sovran (1978)]. The ratio of the drag coefficient to that at zero yaw angle has been used here, to eliminate the problems of definition associated with absolute values of drag coefficient. Furthermore, the increase in drag of 55 per cent is typical for a range of commercial and private vehicles. The drag increases with a non-zero yaw angle, because of the way that the flow

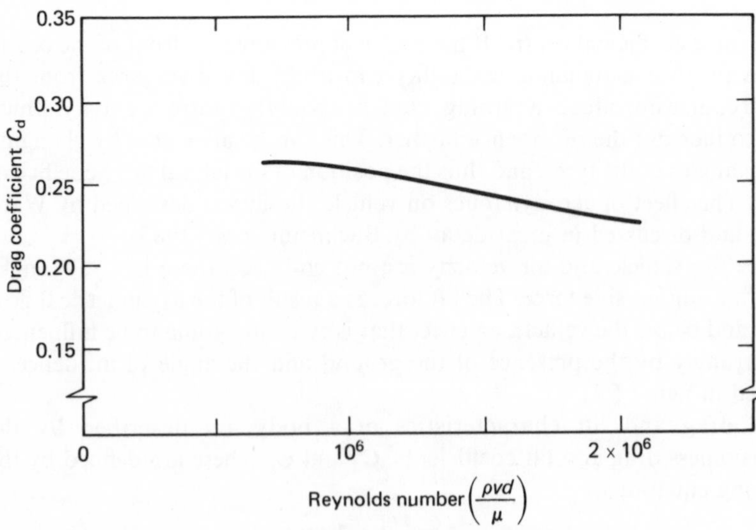

Figure 5.2 Effect of Reynolds Number on the drag coefficient, after Hucho (1978)

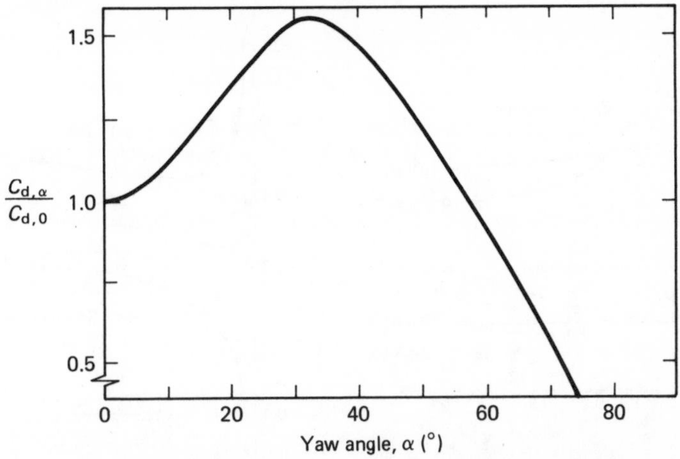

Figure 5.3 The effect of yaw angle on drag coefficient, Sovran (1978). [With acknowledgement to the Plenum Publishing Corporation]

will separate from the side of the vehicle. Since, in general, there will be a wind that is not in the direction of the vehicle motion, the sensitivity of the drag coefficient to yaw is very important. Indeed, a reduction in zero yaw drag at the expense of the peak drag occurring at a lower yaw angle is undesirable. Conversely, an increase in zero yaw drag may be beneficial if it reduces the maximum drag coefficient.

Any tests that are designed to reveal the true drag and lift forces must take into account the ground effect, and the only way that this can be properly modelled is by having a moving ground plane. Bearman (1978) describes a series of experiments on an idealised vehicle model in which the ground clearance (between both a stationary and a moving ground) was varied; the results are shown in figure 5.4. These results show (as has already been stated) that the effect of the ground is more pronounced on the lift forces than the drag forces. The lift coefficient is very sensitive to the angle of incidence (β), especially at the low ground clearances that are found in automotive applications. In comparison, large ground clearances, the angle of incidence and the ground motion all have a comparatively small effect on the drag. For both lift and drag, the effect of the moving ground is to decrease the effective angle of incidence, thus increasing the down-force (negative lift) and increasing the drag.

The higher velocity of the flow over the vehicle roof results in a lower pressure than under the vehicle body, where the flow velocity is low. According to potential flow theory, which can be used to describe the flow outside the boundary layer, the pressure difference between the topside and underside of the vehicle leads to circulation and a lift force. Furthermore, the presence

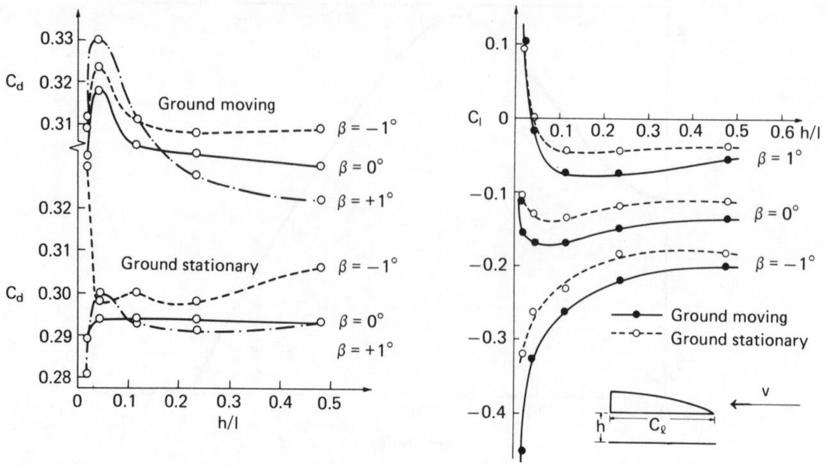

Figure 5.4 The effect of ground clearance, angle of incidence and ground motion on the drag and lift coefficients, Bearman (1978). [With acknowledgement to the Plenum Publishing Corporation]

of circulation implies vorticity, and since vorticity has to be preserved, there will be two trailing vortices as shown in figure 5.5. The inter-relation between lift and drag is highly complex, suffice to say that there are examples of modifications decreasing both lift and drag, increasing both lift and drag, and increasing drag while decreasing lift. Theoretically, the minimum drag would be expected to occur with zero lift.

Decreasing the lift force is of great importance in racing cars, where negative lift (or down-force) is produced at the expense of increased drag. Clearly, the maximum speed is reduced, but the high down-force enables much greater

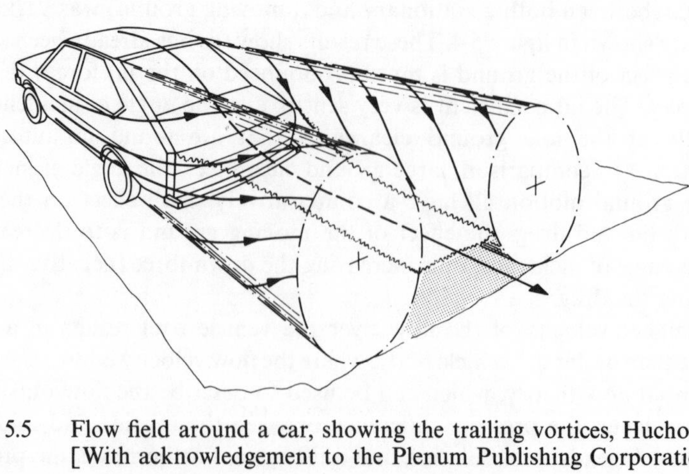

Figure 5.5 Flow field around a car, showing the trailing vortices, Hucho (1978). [With acknowledgement to the Plenum Publishing Corporation]

cornering speeds. Dominy and Dominy (1984) show how down-force is produced on a racing car (figure 5.6), and they state that the down-force can be three times the vehicle weight at a speed of 270 km/h (170 mph). The down-force arises from the inverted aerofoil (which obviously increases the drag), and the low-pressure region produced by the diffuser-like geometry under the body on either side of the driver. To minimise the inflow of air from the sides, it is necessary to use flexible skirts. As with more conventional vehicles, a down-force is produced by the low-pressure region behind the vehicle 'feeding' under the body. Since this low-pressure region is one of the sources of drag, it is clear that the down-force can only be increased at the expense of greater drag.

In order to understand the methods used to reduce drag, it will first be necessary to discuss the mechanisms that produce drag, and how these contribute to the total drag force. The theory will only be outlined here, since this will be sufficient to provide clear definitions of the terms that are used; much fuller treatments can be found in many engineering fluid mechanics books–for example, Massey (1983).

When a fluid flows over a surface at constant speed, a drag force will be produced that consists of two parts: skin friction drag (D_f) caused by viscous effects at the surface, and pressure drag (D_p) as a result of the pressure distribution from the main flow (including the wake) acting on the body surface. The flow over part of a surface is shown in figure 5.7, along with the resultant pressure distribution. Consider area dA at point P; the component

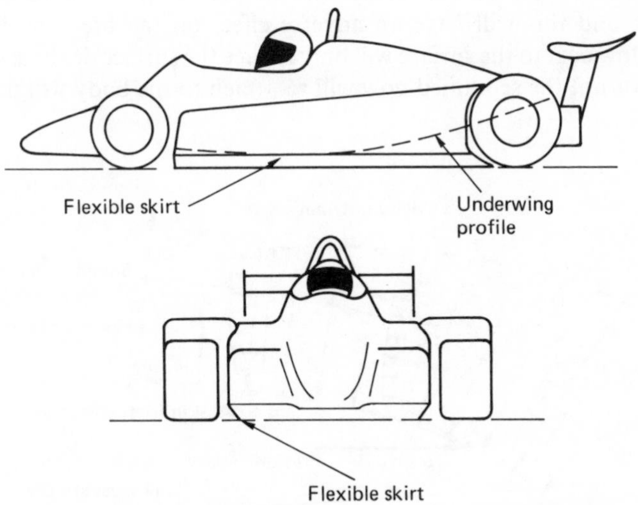

Figure 5.6 Diagram of a modern Formula 1 racing car illustrating the flexible skirts and the side wing profile. Dominy and Dominy (1984). [Reprinted by permission of the Council of the Institution of Mechanical Engineers]

of drag due to the pressure distribution is p sin dA, and the component of drag due to the skin friction is τ_w cos dA. Thus for a complete body the total drag is

$$D = D_f + D_p = \int_A \tau_w \cos \phi \, dA + \int_A p \sin \phi \, dA \qquad (5.2)$$

where the wall shear stress

$$\tau_w = \mu \left(\frac{du}{dy} \right)_w$$

In theory, the pressure distribution can be found by assuming inviscid flow outside the boundary layer and solving the potential flow equations; the shape of the boundary layer and the velocity distribution can be described by empirical correlations. In practice, separation occurs (that is, the flow does not adhere to the surface), and since the position of separation and the nature of the subsequent flow are difficult to predict, complete numerical solutions are not possible.

Separation occurs where there is a rapid change in the surface direction, or where the pressure is increasing in the direction of the flow (a positive pressure gradient); this is illustrated in figure 5.8. The positive pressure gradient tends to reverse the direction of flow, and this is most significant at the base of the boundary layer where the fluid momentum is smallest. Reverse flow will occur where the velocity gradient away from the wall is zero. Separation prevents a further rise in pressure, as can be seen from figure 5.8, and this will have an adverse effect on the pressure drag. The reversed flow next to the surface will only reduce the surface drag very slightly. Whether or not the separated flow will re-attach to the body will depend on

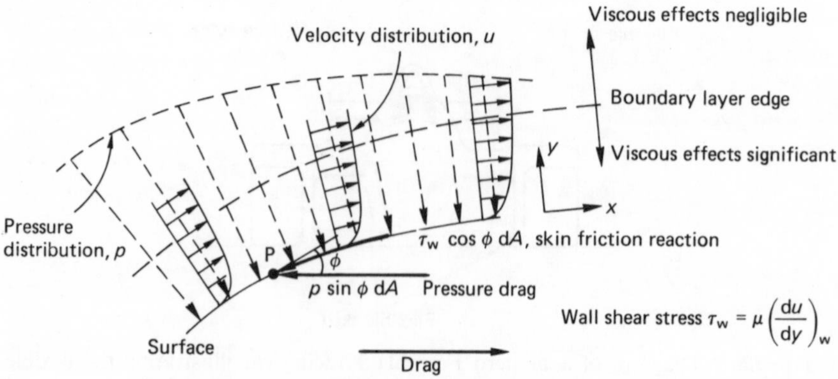

Figure 5.7 The influence of pressure and velocity distributions on drag

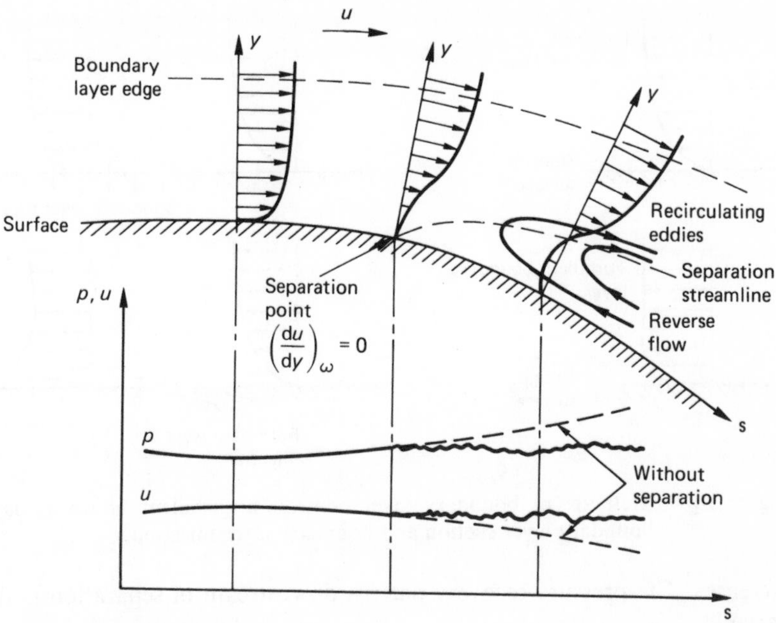

Figure 5.8 The effect of flow separation on the velocity and pressure distribution

the subsequent surface geometry. The reversed flow forms large irregular eddies that dissipate energy from the mainstream by viscous action.

The drag coefficient for a streamlined body with no separation will be about 0.05, and this is almost entirely due to surface drag. For a realistically shaped vehicle body there will be separation, and the lowest drag coefficient feasible is likely to be about 0.15. Since any separation increases the drag profoundly, separation should be reduced even at the expense of increased skin friction drag. Turbulence in the boundary layer increases the skin friction, but since the momentum in the fluid close to the surface is greater, this delays the onset of separation, and can lead to an overall reduction in drag. Other means of boundary layer control are possible; some examples are boundary layer suction or flow injection, and these are shown in figure 5.9. Obviously, any gains in drag reduction must be balanced against the energy cost associated with providing the drag reduction. However, there is scope for using ventilation or cooling flows in this way.

It was stated at the beginning of this section that the centre of pressure tends to move forwards as the vehicle speed increases. The centre of pressure will depend on a summation of the dynamic pressure terms, which are a function of velocity squared, and on the nature of the separated flow. The points of separation in the external flow are essentially fixed, but the positions of flow re-attachment will tend to move back along the vehicle as the speed

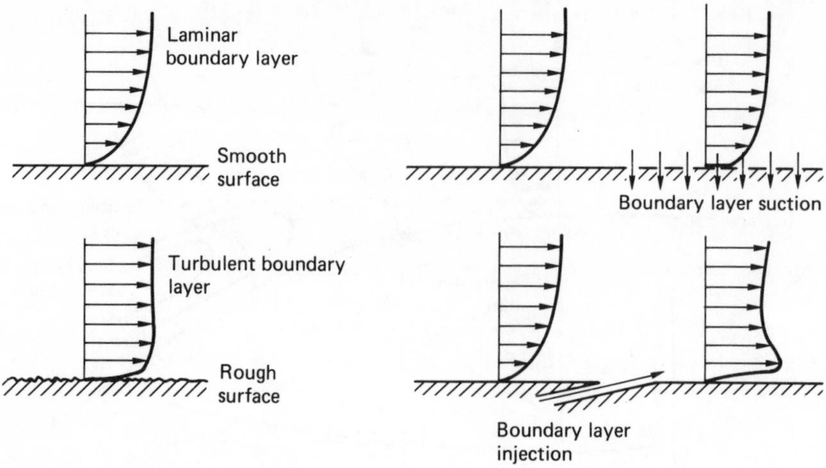

Figure 5.9 Methods of boundary layer control: a turbulent boundary layer, boundary layer suction and boundary layer injection

increases. The pressure recovery pattern downstream of separation is thus variable.

For an actual vehicle, it is difficult to apportion the source of drag between the skin friction drag and the pressure drag; this is especially true for the flow under the vehicle. Are the wheels, transmission and suspension elements to be treated as rough surfaces that contribute to skin friction, or are they to be treated as bodies that contribute to the pressure drag? An approximate breakdown of drag for cars is given in table 5.1.

Table 5.1 Breakdown of the contributions to car drag coefficients

	Rectangular three box saloon, typical of the 1970s	Streamlined hatchback, typical of of the 1980s
Idealised vehicle shape	0.25	0.15
Vehicle with wheels, transmission and suspension	0.35	0.25
With air flow to radiator	0.40	0.27
With surface irregularities caused by body trim, doors and glass	0.45	0.30

The trailing vortices shown in figure 5.5 are a result of the circulation around the car which also causes the lift forces. The vortices obviously contain kinetic energy, and this is obtained from the mainstream, as part of the work in overcoming the drag force. This component is termed 'induced drag' and forms part of the pressure drag. In aerofoil theory the induced drag can be related to lift, but the flow around a vehicle is much more complex. None the less, reducing the lift force to zero should minimise the induced drag.

Another part of the pressure drag arises from what is called 'internal drag'. Internal drag arises from the loss of momentum in the flows that are used for cooling and ventilation. The way that these flows exit from the vehicle also needs to be considered, as they can have either an adverse or beneficial effect on both the skin friction and the pressure drag. The flow through the radiator is an order of magnitude larger than any other flow, so it is the only one discussed here. Consider a flow rate of air $\dot{Q}_r$ to the radiator area A_r. If the vehicle velocity is v, and the momentum from the flow is entirely dissipated, then the drag force, D_r, is given by

$$D_r = \rho \dot{Q}_r v \tag{5.3}$$

and the radiator drag coefficient, C_{dr}, is

$$C_{dr} = \frac{D_r}{\frac{1}{2}\rho v^2 A} = \frac{\rho \dot{Q}_r v}{\frac{1}{2}\rho v^2 A}$$

If $\dot{Q}_r = A_r v_r$, where v_r is the flow velocity into the radiator, then

$$C_{dr} = \frac{2 v_r A_r}{v A} \tag{5.4}$$

For typical values of v_r/v and A_r/A, Hucho (1978) reports that

$$0.01 < C_{dr} < 0.06$$

5.2.2 Experimental techniques

The drag and lift coefficients, and any other information about the flow or pressure distribution around a vehicle, have to be determined experimentally. The most controlled conditions will occur in wind tunnels, but these do also have to be representative of atmospheric conditions.

Since the cost of wind tunnels increases with size, most experimental work is conducted with models. As all the features from a full-sized vehicle can neither be copied on to a model nor adequately scaled down (for example, surface roughness), these will be one source of the discrepancies between the model and full-size tests. The presence of an object in the wind tunnel is also going to modify the flow in the tunnel. Where the cross-sectional area of the model is only a few per cent of the working section of the tunnel, then the effect of the blockage can be neglected. Also in a wind tunnel, there will be boundary layers on the tunnel walls, and these will also influence the mainstream flow. The effects of boundary layers and working section blockage can be allowed for, and the methods are described in books on wind tunnel testing techniques, such as Pankhurst and Holder (1952).

Already there are two conflicting requirements: firstly, that the model should be large in order to give more representative results, and secondly, that the model should be small so as to minimise the blockage effects. In addition, the flow should have dynamic similarity. To give the same ratio of inertia to viscous friction forces, the Reynolds numbers (Re) need to be the same, as given by

$$Re = \frac{\rho v d}{\mu} \tag{5.5}$$

where ρ = fluid density
v = flow velocity
d = characteristic body dimension
μ = dynamic viscosity.

If the wind tunnel is operating with air at atmospheric conditions, then a quarter scale model will require flows at four times the full-scale speed. While the drag coefficient is fairly insensitive to the Reynolds number (figure 5.2), the position of any flow separation is likely to vary, and this will change the position of the centre of pressure.

If a vehicle is designed to travel at 135 km/h, then the Mach number (M) is 0.1, and the effects of air compressibility are negligible. (Mach number is the ratio of air velocity to the velocity of sound, $v/\sqrt{(\gamma RT)}$.)

For a quarter scale model at the same Reynolds number, the Mach number would be 0.4, and the effects of compressibility would be significant. This can necessitate testing at the scale of twelve inches to the foot (that is, full size), or using pressurised wind tunnels to increase the air density .

A typical wind tunnel is shown in figure 5.10. The aim is to produce a uniform flow with a low level of turbulence (local, small-scale velocity fluctuations). Return circuits are used on all but the smallest wind tunnels, since the kinetic energy of the air is preserved, and the power input is thus minimised. There are two reasons for accelerating and then decelerating the flow. Firstly, by placing the fan in the slow-speed section, the power input is minimised. Secondly, the contraction accelerates the main stream without changing the scale of the turbulence; the significance of the turbulence introduced by the fan is thus reduced. Downstream of the working or test section of the tunnel it may be open to the atmosphere, if ambient conditions are to prevail in the test section.

The model should be mounted in such a way that permits the forces and moments to be measured by a balance; the balance is usually mounted outside the tunnel. If necessary, corrections need to be applied to allow for the loading

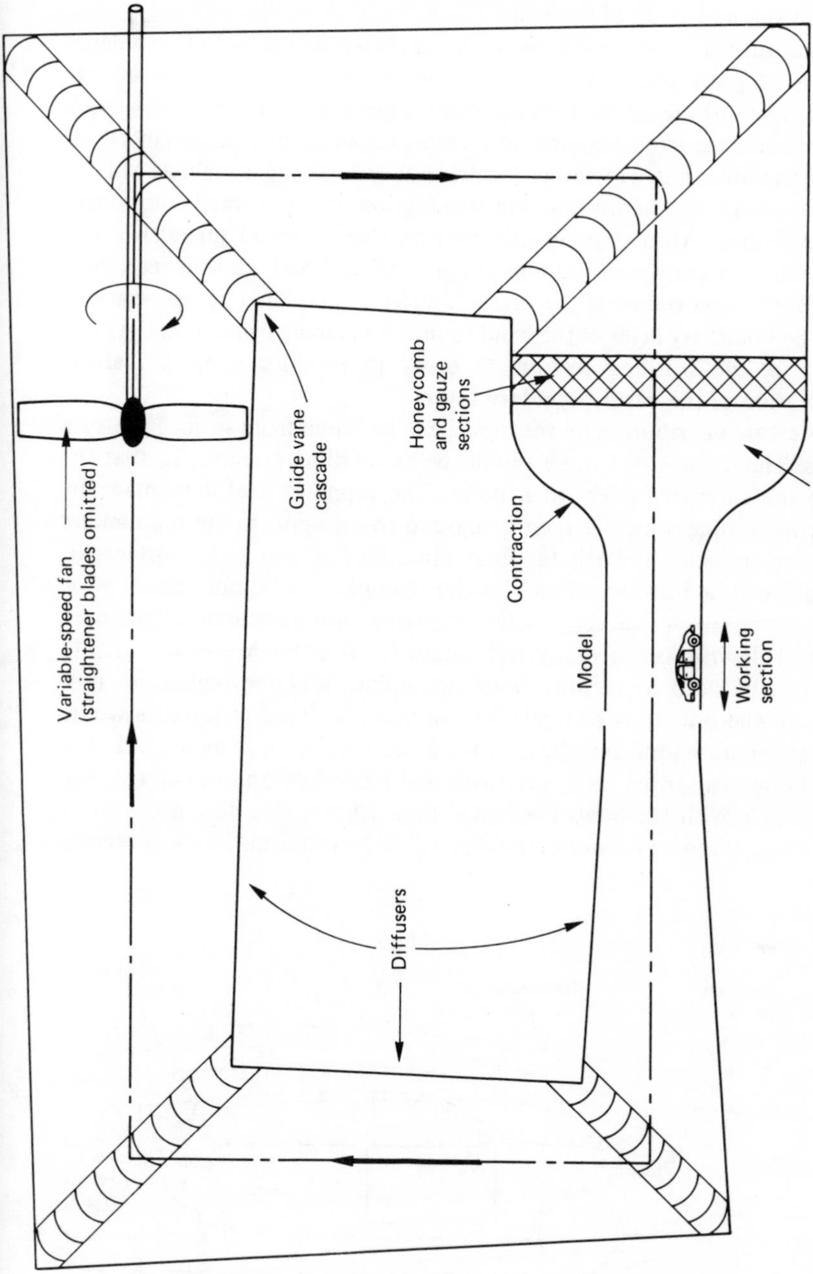

Figure 5.10 A return flow wind tunnel

on the model support. To investigate the ground effect a moving belt is needed. To prevent the belt surface lifting up, it may be necessary to apply suction to the underside of the belt.

If the turbulence level is made too low, it is then possible to add turbulence by means of grids and screens, so that the correct turbulence levels are obtained. The turbulence level is important, since a turbulent flow will cause an earlier transition from a laminar to turbulent boundary layer; this increases the skin friction drag. However, the boundary layer on a vehicle will be mostly turbulent, and in any case the skin friction drag is a small component of the total drag. More significantly though, the increased turbulence will tend to delay separation of the flow (figures 5.8 and 5.9), and this can have a notable effect on reducing the pressure drag. It may also be necessary to modify the boundary layer of the wind tunnel, by placing objects on the floor upstream of the working section, in order to produce a more realistic representation of the earth boundary layer.

The pressure distribution on the model can be found from surface pressure tappings. The diameter of these should be as small as possible, so that the pressure measurement refers to a point. The tappings are often made by hypodermic tubing which is then connected to an appropriate manometer or pressure transducer. Such tappings must be flush with the surface; or turbulence will be introduced and invalid readings will be obtained.

Flow visualisation is a very useful technique with vehicle aerodynamics; a practical description of many techniques is given by Merzkirch (1974). Smoke can be used to identify both streamlines and the regions of flow separation where there is recirculation. In practice, these techniques would be used in separate tests, but the combined effect is shown in figure 5.11. The smoke is often vaporised oil or kerosene, and its density and velocity should be compatible with the flow. The 'rake' that delivers this flow needs to be streamlined, otherwise flow disturbances will be introduced. To show regions

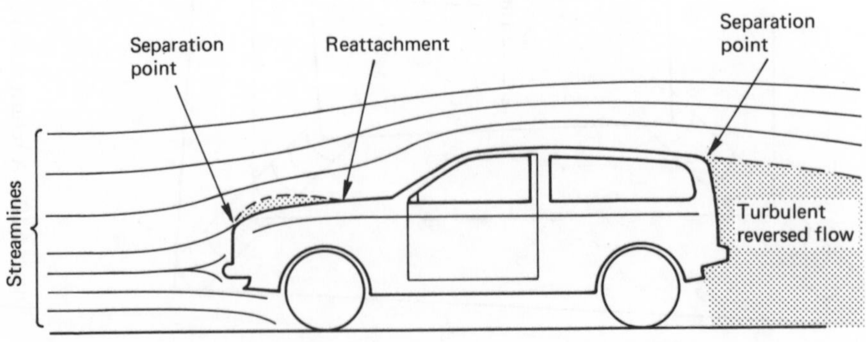

Figure 5.11 The use of smoke to show: (i) streamlines, (ii) regions of separation

of recirculation, the smoke can be admitted by a tubular 'wand'; this can also be used to identify single streamlines.

Tufts of wool or other fibres can be glued to the model, and these show the local surface flows; this is useful when trying to predict water and dirt deposition patterns. Tufts can also be mounted on a grid downstream of the model, in order to show the flow structure in the wake. Surface flow effects of a longer time-scale can be investigated by covering the model with oil or applying the oil in discrete dots. Sometimes pigments are added to the oil–for example, lamp black or titanium dioxide. Photography is an important technique with all forms of flow visualisation. Different types of information can be recorded by using short exposures, long exposures or ciné film; experience and experiment are both usually necessary. Occasionally, experiments are conducted with models in water flows; this can facilitate some aspects of flow visualisation.

An alternative approach for finding drag coefficients is to conduct coast-down tests on full-size vehicles. This approach also provides information on the vehicles rolling resistance, but the experimental difficulties can be significant. The tests need to be conducted on a straight road of known inclination (preferably constant or zero), under windless conditions. Even wind speeds that are low compared with vehicle speeds produce a flow with yaw, and it has already been shown that the drag coefficient is very sensitive to the yaw angle. Also, even if there is a wind with constant velocity, the yaw angle will change as the vehicle decelerates. The forces on the vehicle control the deceleration as follows:

$$M_{\text{eff}}\frac{\mathrm{d}v}{\mathrm{d}t} + R + \frac{1}{2}\rho v^2 A C_{\mathrm{d}} + Mg \sin\theta + U = 0 \qquad (5.6)$$

where

$$M_{\text{eff}} = M + \frac{I}{r^2}$$

and M = vehicle mass
I = inertia of the roadwheels and drivetrain referred to the wheel axis
r = wheel radius
R = rolling resistance (assumed to be independent of speed)
θ = inclination of the road–upwards in the direction of travel taken as positive
U = unsteady flow term, which is negligible in coast-down tests

The experiment is likely to record position or speed as a function of time, rather than deceleration as a function of velocity. Thus the data have to be differentiated or equation 5.6 has to be integrated. Both these approaches are discussed by Evans and Zemroch (1984), along with a method of fitting the data to equation 5.6, in order to determine the drag coefficient and rolling resistance by linear regression. The tests need to be planned carefully, with coast-down from a range of vehicle speeds in both directions.

5.3 Automobile aerodynamics

5.3.1 The significance of aerodynamic drag

The general aerodynamic considerations have already been discussed in the previous section, so the two key aspects in this section are how reductions in automobile drag coefficients are obtained, and what effect they have on fuel economy.

First it will be useful to make a comparison between the rolling resistance and the aerodynamic resistance, and these results can be seen in figure 5.12. A constant rolling resistance of 225 N is assumed, and the aerodynamic resistance (drag) has been plotted for two cases: $C_d = 0.33$, and $C_d = 0.45$.

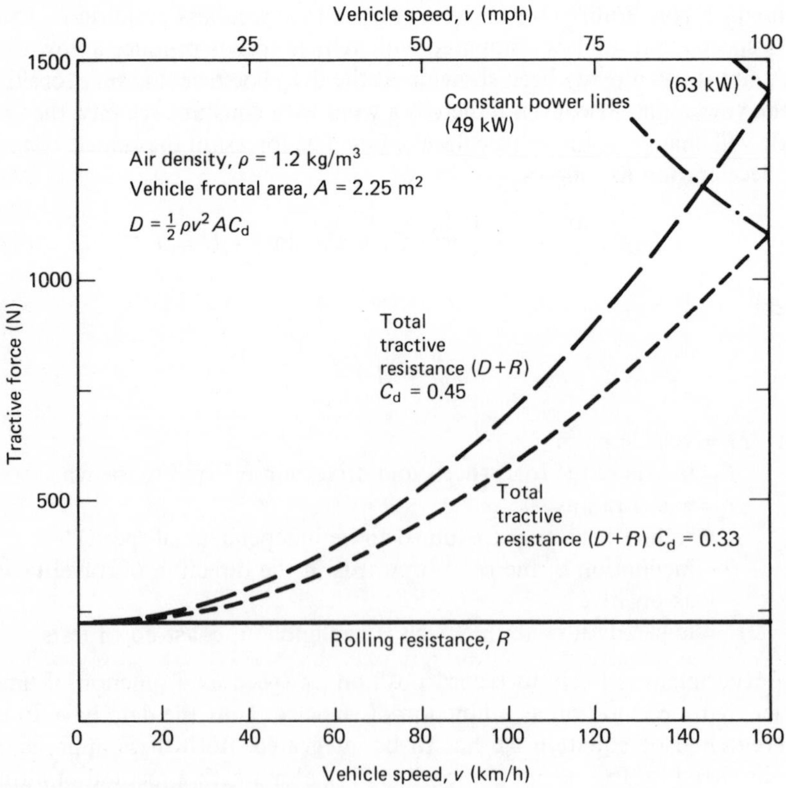

Figure 5.12 The effect of aerodynamic drag on automobile performance

In each case the frontal area is assumed to be 2.25 m^2. These values are typical of a medium-sized saloon of the mid-1980s and mid-1970s respectively. Since aerodynamic resistance is proportional to speed squared, the resistance is insignificant at low speeds, but then increases rapidly and becomes very significant at high speeds. The aerodynamic resistance (drag, D) equals the rolling resistance at 80 km/h and 70 km/h for the cases where $C_d = 0.33$ and 0.45 respectively. Thus, it is at the higher speeds that the reductions in drag will have the greatest effect on the automobile performance; this is most evident for the maximum speed. The power (W) required to propel a vehicle is the product of the tractive force (N) and speed (m/s); the constant power lines are thus rectangular hyperbolae on figure 5.12. Thus, for a given power (49 kW), the reduction in drag from $C_d = 0.45$ to $C_d = 0.33$ will allow an increase in maximum speed from 145 to 160 km/h. Alternatively, if the vehicle with $C_d = 0.45$ is required to travel at 160 km/h, then 63 kW will be required. It is important to remember that the power used in overcoming aerodynamic drag is proportional to the speed cubed.

The effect of aerodynamic drag reductions on vehicle acceleration will be small. The difference between the total tractive resistance and the tractive force available from the powertrain is used to accelerate the vehicle. This difference will reduce as the speed increases, since the tractive resistance increases and the available tractive force (F) reduces (assuming constant power). Consider the acceleration from 0 to 100 km/h. The maximum possible tractive force available at 100 km/h with 49 kW maximum power is 1765 N, while the tractive resistance is 694 N; this leaves a balance of 1071 N for acceleration. Reducing the aerodynamic drag coefficient to $C_d = 0.33$ increases the force available for acceleration at 100 km/h by 11.7 per cent, and this represents an upper bound on the reduction in acceleration time, since at lower speeds the reduction in drag will be even less significant. This case is considered further in example 5.1 of section 5.7, where the equations of motion are solved; this exact analysis shows a 4.2 per cent reduction in the acceleration time.

The effect of the reduced aerodynamic drag on fuel economy will evidently be most significant at the highest vehicle speeds. A simple argument would suggest that a 10 per cent reduction in total tractive resistance would give a 10 per cent reduction in fuel economy. The supposition here is that the powertrain efficiency remains the same, as a result of changing the transmission ratios and/or reducing the size of the engine. Nor is any account taken here of the second-order effects; for example, reducing the size of the engine gives a weight saving throughout all the powertrain components, and a weight reduction will reduce the rolling resistance. On this simple basis, the potential fuel savings are shown in table 5.2.

These fuel savings are not achieved in practice, since vehicles are not driven at constant speeds. The requirements to model the fuel consumption of a vehicle were established in section 4.6; in particular the following steps are

Table 5.2 The effect of reducing the aerodynamic drag from $C_d = 0.45$ *to* $C_d = 0.33$ *for constant speed fuel consumption* ($A = 2.25\ m^2$, $\rho = 1.2\ kg/m^3$, $R = 225\ N$)

Speed km/h	50	80	120	160
Reduction in fuel consumption	6.7%	15.2%	20.0%	22.5%

necessary:

(a) The driving pattern needs to be defined (speed as a function of time).
(b) The powertrain efficiency has to be defined.
(c) The aerodynamic and rolling resistance need to be defined.

Hucho (1978) has demonstrated that correlations exist between frontal areas, mass and power for European cars, and this enables a widely applicable fuel economy model to be developed. Hucho assumes a given engine efficiency map and a fixed transmission efficiency of 90 per cent; the fuel consumption is a weighted average from the highway and urban driving cycles. The results from this model are shown in figure 5.13 for three different sized vehicles. Figure 5.13 suggests that a reduction in drag coefficient from 0.45 to 0.33 would lead to an overall gain in fuel economy of 9 per cent.

5.3.2 Factors influencing the aerodynamic drag

The aerodynamic drag of a vehicle will depend on both the overall shape of the vehicle (whether it is, for example, a notchback or a hatchback), and the

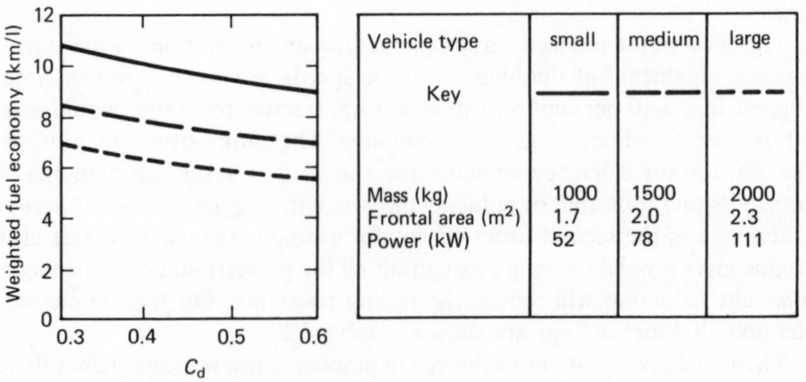

Figure 5.13 The effect of drag coefficient on vehicle fuel economy, adapted from Hucho (1978)

body details such as the gutters at the edge of the windscreen or the wheel trim. Despite the apparent dissimilarities between vehicles, if cars are grouped by size into small, medium and large sizes, and by body type as either notchback or hatchback, then in each group the centreline cross-sections and wheelbase sizes are remarkably similar [Hucho (1978)]. None the less, in each category there is a significant variation in drag coefficient, and this has to be attributed to differences in detail design.

In hatchback cars, the angle of inclination of the rear window determines whether separation occurs at the top or bottom of the rear window; naturally this has a strong influence on the drag coefficient as shown by figure 5.14. Evidently, when separation occurs below the rear window, dirt deposition on the rear window will be a less serious problem. The height of cars is almost independent of size, and the height of the bottom edge of the rear window is dictated by visibility requirements. Thus, longer cars can accommodate smaller angles of inclination for the rear window, which leads to lower drag coefficients.

The nose geometry can also be very important through its influence on the position of flow separation on the bonnet (hood), as shown by figure 5.15. This is apart from the arrangement of the air flow into the engine compartment, that has been discussed in section 5.2.1. Hucho *et al.* (1976) also give results for drag reductions as a result of many minor design changes; for example

(a) Rounding the transition from the roof to the rear windows can give a 9 per cent drag reduction.

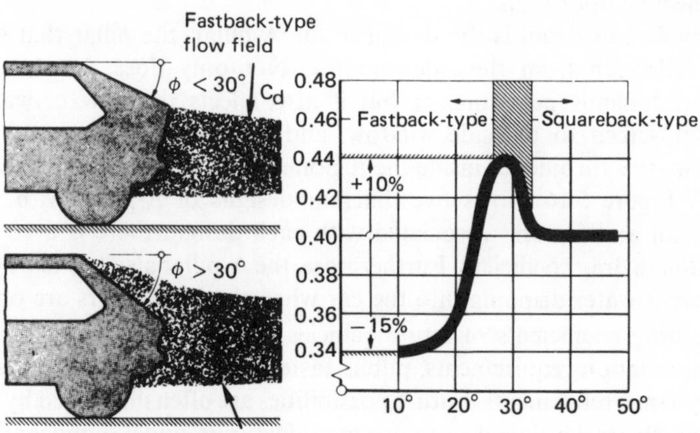

Figure 5.14 The influence of rear window inclination on the drag coefficient, C_d, and the region of separation, Hucho *et al.* (1976). [Reprinted with permission © 1976 Society of Automotive Engineers, Inc.]

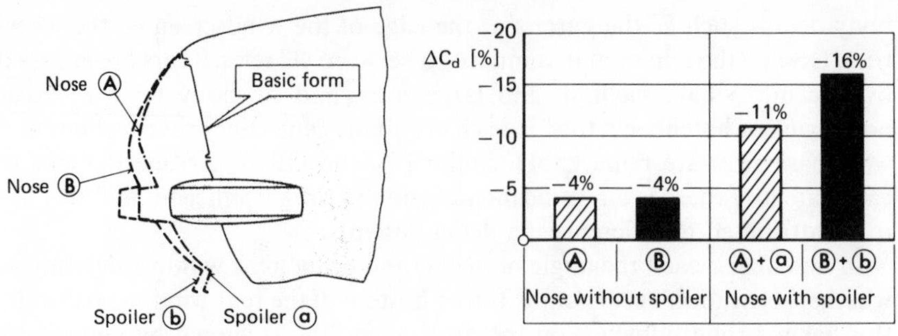

Figure 5.15 The influence of nose geometry on the drag coefficient, from Hucho *et al.* (1976). [Reprinted with permission © 1976 Society of Automotive Engineers, Inc.]

(b) Reducing the width of the car to the rear can give a 13 per cent drag reduction.

Also, by ensuring that separation does not occur over the bonnet and windscreen, not only is the drag coefficient minimised, but also the pressure rise at the base of the windscreen is increased. This pressure rise is important, since it is the source of ventilation for the passenger compartment.

Attention to detail is vital if the drag coefficient of a streamlined basic body shape is not to be increased inordinately. Typical of the approach that leads to a low drag coefficient is the use of flush mounted glass, recessed windscreen wipers, optimised rear view mirrors, low turbulence wheel trims and effective door seals.

A significant detail is the design of the A-pillar, the pillar that separates the windscreen from the side window. Not only does the design affect the aerodynamic performance, but it also affects the flow of water from the windscreen to the side window, and the turbulence around the side window–this turbulence inside the region of flow separation produces 'wind noise'. Figure 5.16 shows five different designs of A-pillar with the drag coefficient and features associated with each design: there is a 10 per cent variation in drag coefficient. Furthermore, the A-pillar also has to be designed to prevent water dripping into the car when the front doors are opened.

The drag coefficients of future vehicles will be governed by the interior accommodation requirements, public taste and the cost effectiveness of any aerodynamic refinements. Future possibilities are often illustrated by 'concept cars' which are developed by motor manufacturers, and a significant example of this type is the Ford Probe IV vehicle that is shown in figure 5.17.

The design specification was for a fully functional vehicle with seating for four passengers and a drag coefficient below 0.2; the vehicle is described by Santer and Gleason (1983) and Peterson and Holka (1983). Many radical

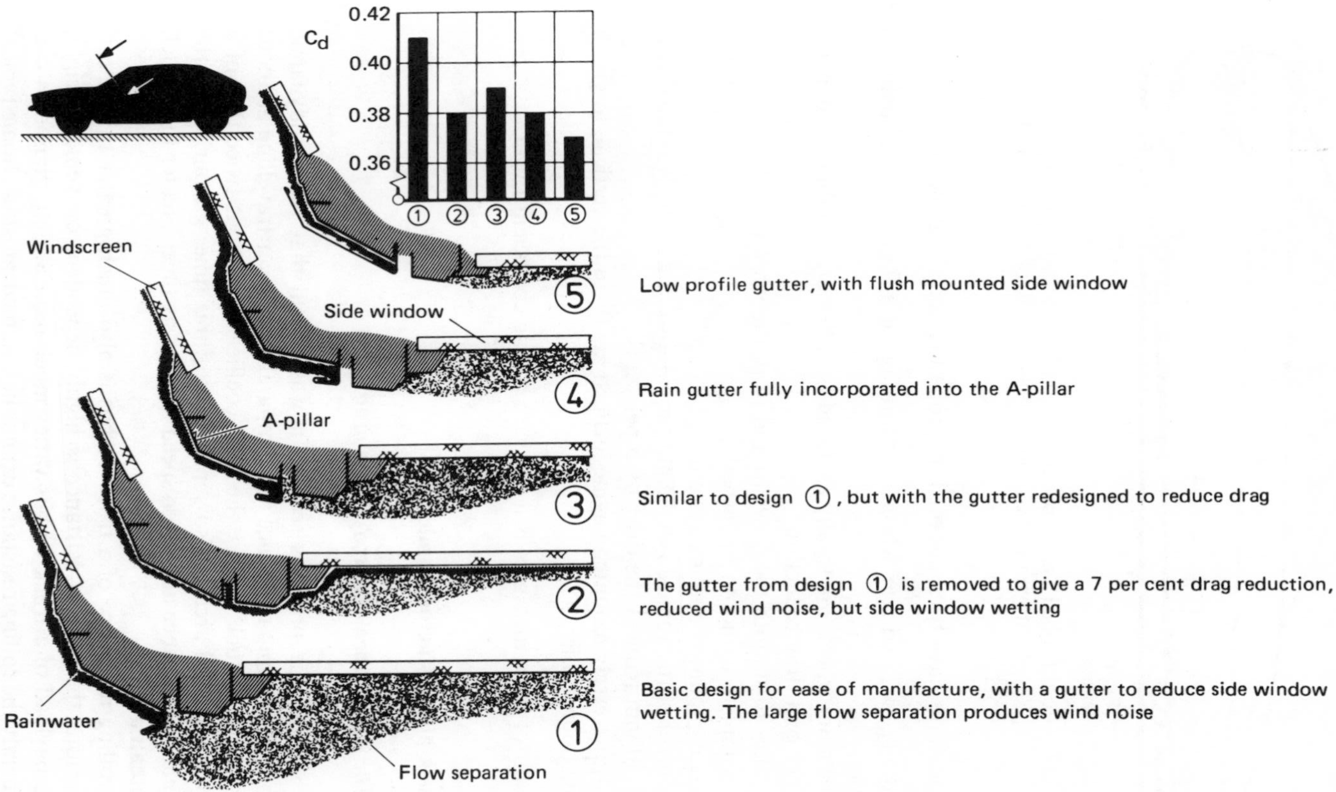
C_d
0.42
0.40
0.38
0.36
① ② ③ ④ ⑤
Windscreen
Side window
A-pillar
Rainwater
Flow separation
⑤ Low profile gutter, with flush mounted side window
④ Rain gutter fully incorporated into the A-pillar
③ Similar to design ①, but with the gutter redesigned to reduce drag
② The gutter from design ① is removed to give a 7 per cent drag reduction, reduced wind noise, but side window wetting
① Basic design for ease of manufacture, with a gutter to reduce side window wetting. The large flow separation produces wind noise

Figure 5.16 Design of the A-pillar to obtain low drag, low wind noise and proper flow of rain water, from Hucho *et al.* (1976). [Reprinted with permission © 1976 Society of Automotive Engineers, Inc.]

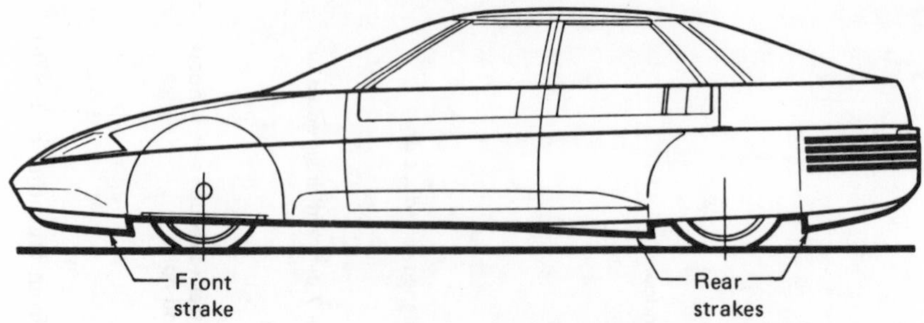

Figure 5.17 Ford Probe IV concept car, $C_d = 0.15$, from Santer and Gleason (1983). [Reprinted with permission © 1983 Society of Automotive Engineers, Inc.]

design features have been adopted in Probe IV, and these include:

(a) Totally enclosed rear wheels, accounting for a 9 per cent reduction in drag.
(b) The front wheels are totally enclosed by a flexible membrane, to give a 5 per cent drag reduction.
(c) Strakes to smooth the flows to and from the wheels.
(d) A completely smooth underbody.
(e) A rear-mounted cooling system.
(f) Control of the vehicle ride height (equal lowering of the front and rear by 30 mm reduces the drag by 5 per cent).
(g) A transversely mounted engine inclined at 70° to the vertical, to give a low bonnet line.

The result of these measures is a car with a drag coefficient of 0.15.

5.4 Truck and bus aerodynamics

5.4.1 The significance of aerodynamic drag

At the start of the previous section, it was shown that the aerodynamic resistance and rolling resistance are comparable for a car travelling at about 75 km/h. For a truck or bus, the drag coefficient is about twice that of a car, and the frontal area is also some two to three times greater. Thus, the aerodynamic resistance at a given speed is some six times greater for a truck or bus than a car.

The rolling resistance of a truck or bus is also much greater than a car. Furthermore, the rolling resistance is highly dependent on vehicle weight, and the payload of a truck or bus varies much more widely than for a car, consequently the rolling resistance cannot be assumed constant. Some typical values of rolling resistance are shown in table 5.3.

Table 5.3 Rolling resistance as a function of mass for different vehicles, a subject discussed further in section 6.2.2

	Car	Truck	
		Unladen	Laden
Mass of vehicle (kg)	1 000	11 000	33 000
Rolling resistance of vehicle (N)	225	1 050	2 250

If a value of $AC_d = 5.7$ m^2 is assumed for the truck in table 5.3, then the variation of aerodynamic resistance with speed will be as shown in figure 5.18. For this particular case, the aerodynamic resistance becomes equal to the rolling resistance at 63 km/h for an unladen vehicle, and at 92 km/h for a laden vehicle. Thus aerodynamic resistance is significant for any vehicle cruising at motorway speeds, and also for unladen or lightly loaded vehicles at significantly lower speeds. Evidently, the aerodynamic design will be particularly significant for high-speed coaches since the payload is small (say 4000 kg for 50 people). Also shown on figure 5.18 is the 100 kW powerline. If this power is available at the rear wheels through appropriate gear ratios, then the laden vehicle can travel at 85 km/h, and the unladen vehicle can travel at 98 km/h.

5.4.2 Factors influencing the aerodynamic drag

The drag coefficient for a rectangular box that is typical of commercial vehicle shapes is about 0.9; this, of course, increases for flows of non-zero yaw angle, and the drag coefficient can become greater than unity. This simple rectangular shape is closest to that of a coach; trucks with integral cabs and bodies and tractor/trailer combinations have radically different shapes, so these will be discussed afterwards.

A coach does not have a purely rectangular shape, and these departures from a simple rectangular shape have a significant effect on drag reduction:

(a) The vehicle front is convex in plan view.
(b) The windscreen is inclined to the vertical.
(c) All corners between the front face and the sides are well rounded.

A well known example of this is the work on the VW Microbus by Moeller (1951) [also reported by Schlichting (1960)]. By making the front of this vehicle convex, with an inclined windscreen, the drag coefficient is reduced to 0.76. However, there is still separation occurring with the flow on to the

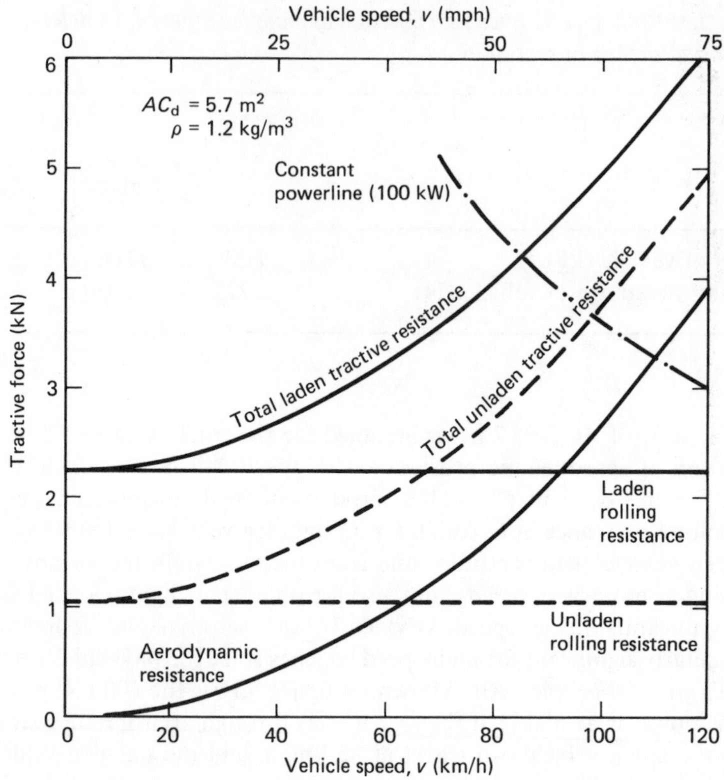

Figure 5.18 The relative significance of aerodynamic and rolling resistances for a truck

top and sides of the vehicle. This separation is eliminated by corner rounding, and the drag coefficient is then reduced to 0.42.

Corner rounding is an important technique in drag reduction, and the radius of curvature is quite small in many cases; typically r/b is approximately 0.1 (where r is the radius of curvature and b is the breadth of the object in the plane of measurement). Since the drag reduction occurs through reducing or preventing flow separation, once the radius of curvature is sufficient to prevent flow separation, any further increase in the radius of curvature will not lead to any further drag reduction. However, the onset of separation occurs earlier with higher Reynolds numbers (effectively flow velocity), and the minimum radius of curvature to prevent flow separation increases with increasing Reynolds numbers; several examples of this can be found in Hucho *et al.* (1976).

There are essentially two types of truck: those built on a single chassis, and tractor/trailer combinations. Tractor/trailer combinations are articulated to improve manoeuvrability, and this necessitates an air gap between the units. Mason and Beebe (1978) report a series of tests in which the spacing between the tractor and trailer was varied. The trailer used had a simple rectangular shape, with no corner rounding; the tractor had the same body width and well rounded corners ($r/b \approx 0.1$). Some significant results are shown in figure 5.19. The drag coefficient of the isolated trailer is 0.92, and the flow is characterised by a separation bubble above the leading edge of the trailer. When the tractor is placed immediately in front of the trailer, the drag coefficient is minimised ($C_d = 0.72$), and this limiting case corresponds to a single chassis truck. As the gap between the tractor and trailer is increased, the drag coefficient increases. Once a gap exists, the flow above the tractor roof divides, and there is some downflow (with associated turbulence) between the tractor and trailer; under these conditions, the drag coefficient will be the sum of the two drag coefficients measured in isolation.

When the tractor is placed immediately ahead of the trailer, the drag reduction is caused by the tractor smoothing the air flow on to the trailer. The drag increase as the separation increases is attributed to a downflow, and this increase in drag can be virtually eliminated by using a horizontal plate to inhibit the downflow; this is illustrated in figure 5.20.

An alternative approach to drag reduction is to minimise the separation between the tractor and trailer. By departing from a simple pivot, the gap between the tractor and trailer can be reduced for a given level of manoeuvrability; some examples can be seen in figure 5.21.

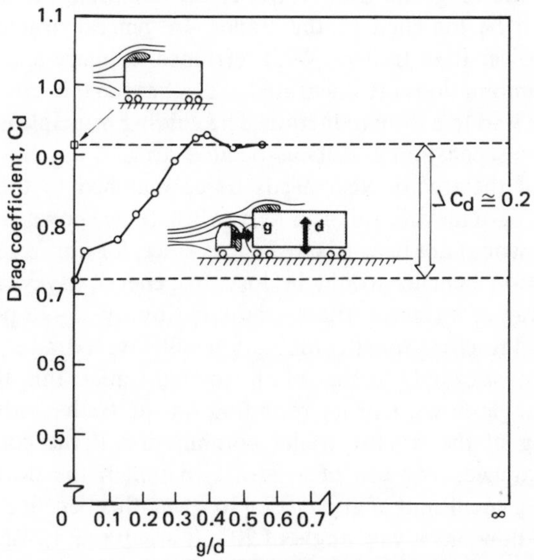

Figure 5.19 The variation in drag coefficient as a function of the tractor/trailer separation, Mason and Beebe (1978). [With acknowledgement to the Plenum Publishing Corporation]

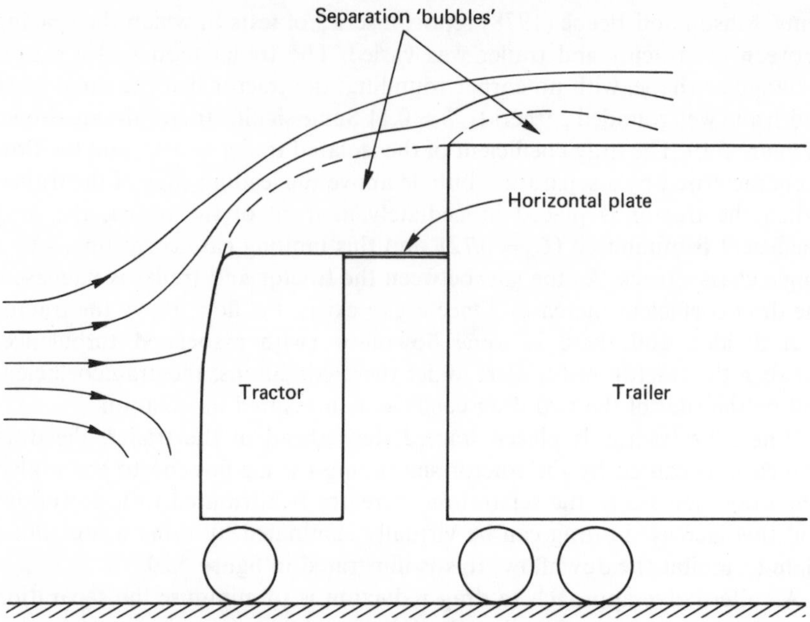

Figure 5.20 The use of a horizontal plate to reduce the drag coefficient by eliminating the downflow between the tractor and trailer

To minimise the drag of a tractor/trailer combination, the flow from the tractor needs to be matched to the trailer. In general, tractors are both narrower and lower than trailers. With reference to figure 5.22, where high and low drag combinations are illustrated, it can be seen that corner rounding does not always lead to a drag reduction. The guiding principle is to minimise the effect of flow separation and its associated drag.

The height of the tractor also needs to be matched to the trailer; the simplest means of achieving this is a vertical plate as shown in figure 5.23. In practice, this would not be very satisfactory since, for non-zero yaw angles, the drag reduction benefits would be lost. In general, a yaw angle of 15° increases the drag of a tractor/trailer combination by 35–45 per cent.

An unmatched tractor/trailer combination will have a drag coefficient that corresponds almost exactly to that of an isolated trailer; this also holds for non-zero yaw angle flows. Corner rounding on the trailer only appears to reduce the drag of the tractor/trailer combination if the combination is otherwise unmatched. The use of a fairing to match the flow can give a reduction in drag coefficient of about 0.2. However, this benefit is lost entirely by the time the flow has a yaw angle of 20°. If a gap seal is used to prevent lateral flow, then the reduction in drag associated with the fairing is maintained.

A significant example of a low drag truck is the General Motors Aero

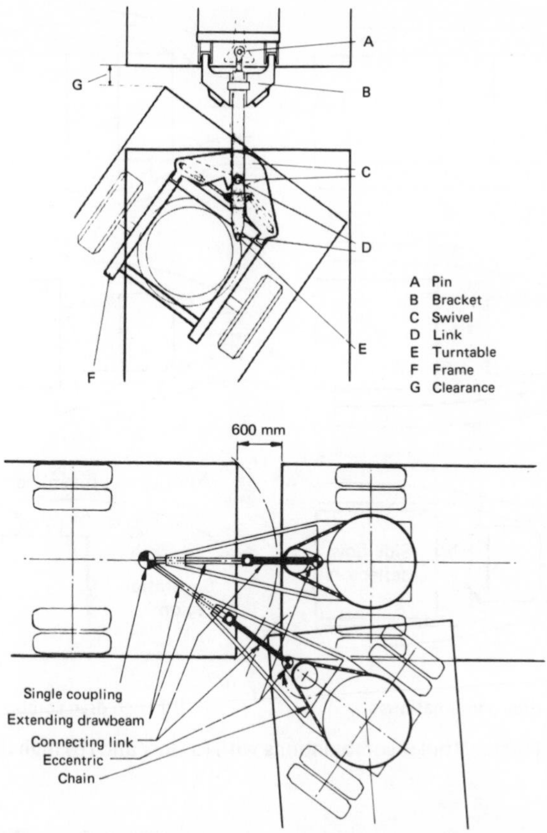

Figure 5.21 Techniques for reducing the tractor/trailer separation. [Reprinted by permission of the Council of the Institution of Mechanical Engineers]

Astro which is described by Gregg (1983). This truck uses a roof-mounted deflector and side gap fillers to produce a drag coefficient in the range 0.4 –0.5.

The other area for improvement is in a reduction of the pressure drag associated with the separation flow behind the vehicle. The most effective device is a vertical plate, and the way this modifies the flow is illustrated in figure 5.24. Mason and Beebe (1978) also report the increase in pressure recovery on the rear face that is caused by the vertical plate. The pressure recovery attributable to the vertical plate leads to a reduction in drag coefficient of 0.04 for a trailer, and 0.09 for a coach. Such a vertical plate might also help to reduce the spray from wet road surfaces.

5.5 Numerical prediction of aerodynamic performance

It should be self-evident that the aerodynamic testing and refinement of

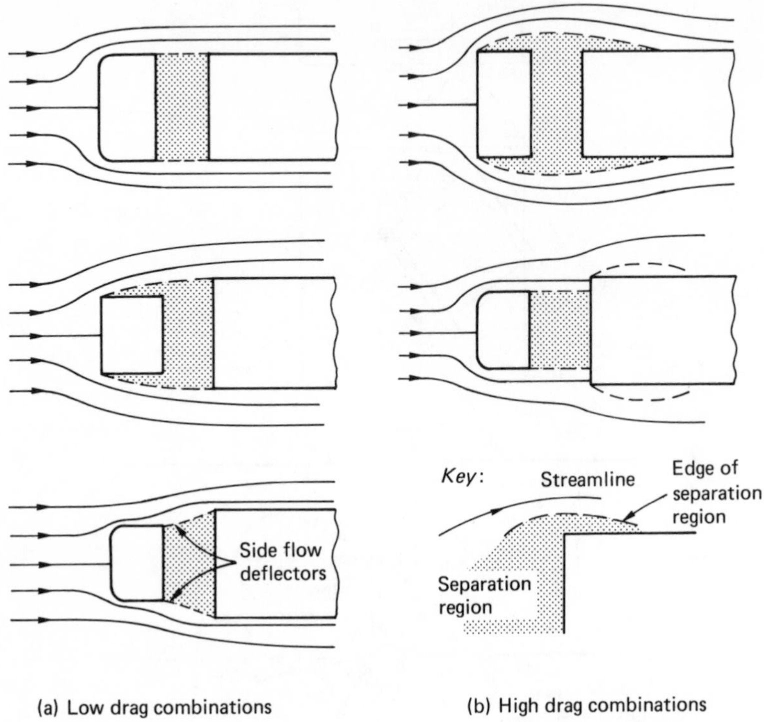

Figure 5.22 Tractor/trailer combinations with (a) low and (b) high drag geometries

vehicle shapes is a time-consuming and expensive activity. The traditional experimental approach can be optimised by drawing on past experience–for instance, the degree of corner rounding, the bonnet and rear window slopes, the extent of windscreen wrap-around, details of glass fixing etc. However, this approach still requires experiments to determine the aerodynamic characteristics of a basic vehicle shape.

An alternative approach is to solve the fluid flow equations numerically, by using the techniques developed for predicting the flow over aerofoils and through turbomachinery. Unfortunately, vehicle geometries are much more complex, since they are three-dimensional non-streamlined bodies adjacent to a surface; consequently, the application of numerical methods to vehicles is more difficult and limited.

The approach in this type of problem is to separate the flow into two parts: the mainstream where viscous effects are negligible, and the boundary layer where viscous effects are significant. The flow in the mainstream is assumed to be subject only to the effects of pressure and momentum (that is, a potential flow), and the fluid flow equations thus simplify to a single linear partial differential equation (Lapace's equation). The boundary layer

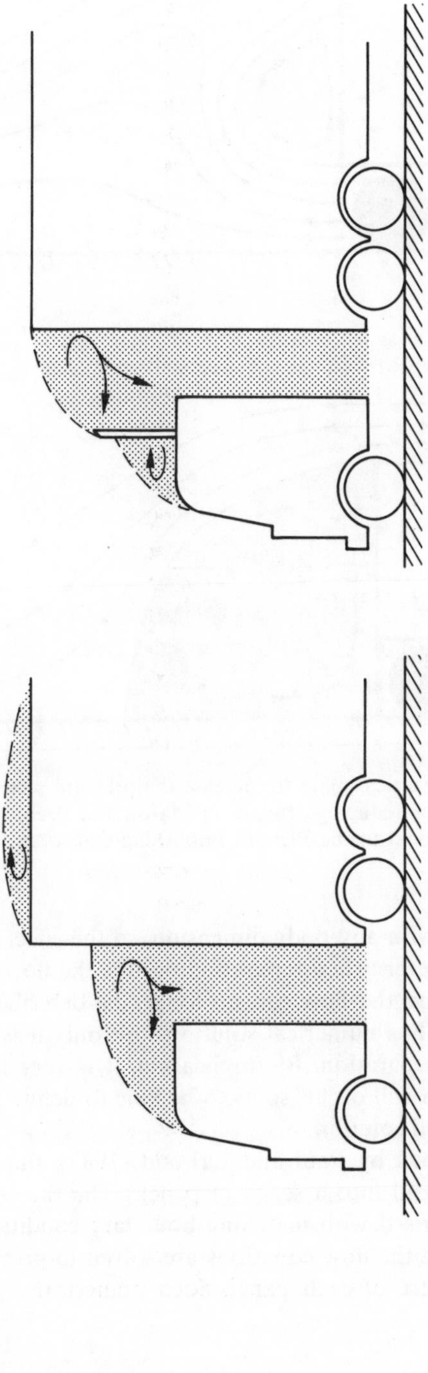

Figure 5.23 The use of a vertical plate to match the air flow from the tractor to the trailer in order to reduce drag

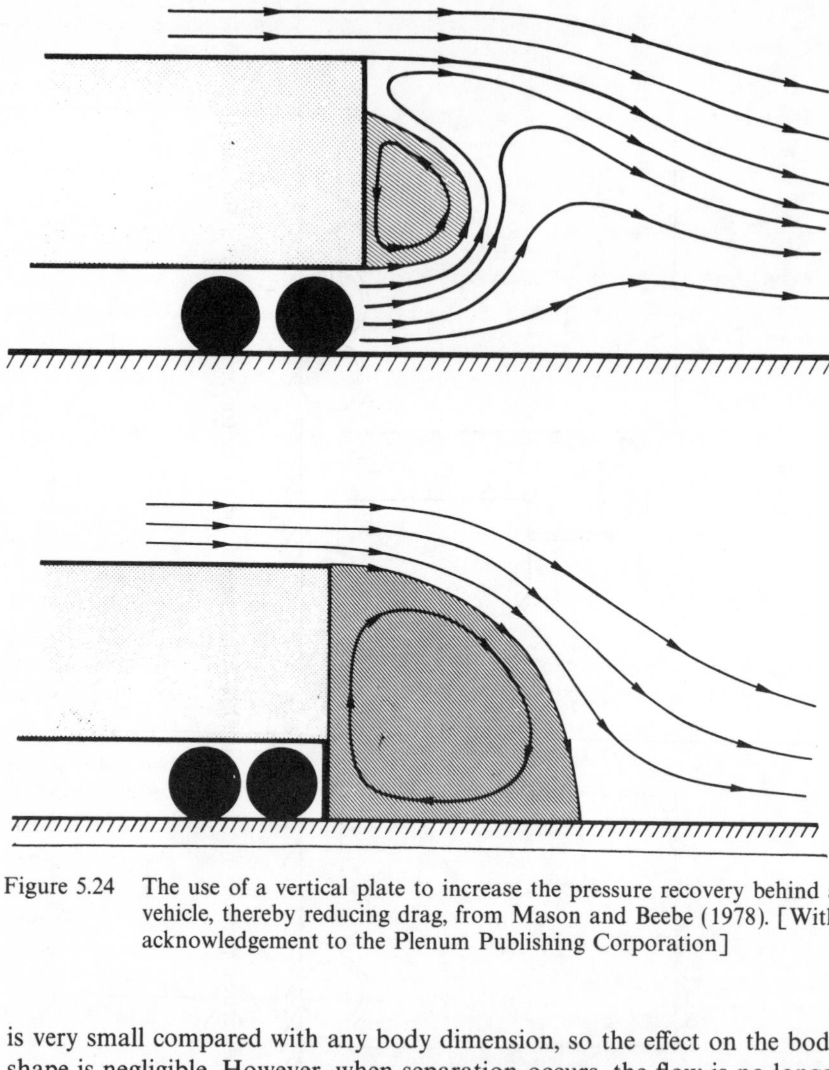

Figure 5.24 The use of a vertical plate to increase the pressure recovery behind a vehicle, thereby reducing drag, from Mason and Beebe (1978). [With acknowledgement to the Plenum Publishing Corporation]

is very small compared with any body dimension, so the effect on the body shape is negligible. However, when separation occurs, the flow is no longer attached to the body, and the flow can no longer be described by simple analytical expressions. Thus numerical solutions are only feasible prior to the occurrence of flow separation. By implication, it is very important to identify where separation will occur, so as to be able to define the extent of validity for any numerical solution.

In the method described by Paul and LaFond (1983), the surface of a vehicle is divided by a grid into a series of panels. The three-dimensional flow equations are linearised with matching boundary conditions between neighbouring panels, and the flow equations are solved to give the velocity and pressure at the centre of each panel. Such numerical results can be checked against experimental results, and against analytical results for very simple geometries. The point of separation is determined by a correlation that considers the down-wind rate of change of static pressure, and the Reynolds number.

5.6 Conclusions

Drag reduction can have a significant effect on the fuel economy of all types of vehicle, especially at higher speeds. The improvements to the steady-state fuel consumption are greater than those attained in normal road use, since negligible energy savings are obtained during low and medium speed acceleration. The effects of reducing drag are a significant increase in a vehicle's maximum speed, but a negligible reduction in most acceleration times.

When considering vehicle drag results, care should be taken in making comparisons between different tests. The drag coefficient will depend on how the area is defined, the nature of any turbulence, the Reynolds number, the velocity distribution of the incident flow, the accuracy of any body details (especially on a reduced scale model), the internal flows and the presence of the ground. It must also be remembered that vehicles are usually subject to a crosswind, and that drag coefficients rise markedly for non-zero yaw angles.

Vehicle aerodynamics is not just concerned with drag reduction. The stability and handling of a vehicle will depend on the position of the centre of pressure and the sign and magnitude of any lift forces. While the ground effect (the relative motion between the vehicle and the ground) has a small effect on drag, it has a profound effect on lift–even to the extent of creating a down-force. The aerodynamic design should also consider the water and dirt deposition patterns on the glass surfaces, wind noise, and the various cooling and ventilation flows.

The three-dimensional flow around a vehicle is highly complex, and complete numerical solutions have not been achieved yet. Consequently, the wind tunnel testing of models and full-size vehicles is still a vital part of the aerodynamic development. Since this is a lengthy and costly process, past experience with techniques such as corner rounding is very important. A key aspect in the aerodynamic development is the control and nature of the flow separation. In this context, flow visualisation is very important, since smoke traces enable streamlines and regions of separation to be identified.

It has been shown here that current drag levels can be reduced still further, with figures as low as $C_d = 0.15$ for a car, and a figure of about three times this for coaches and trucks. However, whether or not such vehicles become commercial propositions will depend on the consumer attitudes towards the appearance, and the additional manufacturing costs.

5.7 Examples

Example 5.1

Calculate the 0–100 km/h acceleration time for a vehicle of effective mass (m) 1000 kg, rolling resistance (R) 225 N for two cases: (a) $C_d = 0.45$, (b) $C_d = 0.33$. As in figure 5.12, $A = 2.25\ \text{m}^2$, $\rho = 1.2\ \text{kg/m}^3$.

Assume that the available tractive force (F) is constant at 3530 N between 0 and 50 km/h, and 1765 N between 50 and 100 km/h.

Solution

Applying Newton's second law to the motion of the car gives

$$F - (R + \tfrac{1}{2}\rho v^2 A C_d) = m\frac{\mathrm{d}v}{\mathrm{d}t} \tag{5.7}$$

$$\frac{\mathrm{d}v}{\mathrm{d}t} = \frac{F - R}{m} - \frac{\rho A\, C_d\, v^2}{2m}$$

This equation needs to be arranged in a form that can be integrated:

$$\frac{\mathrm{d}v}{\dfrac{F-R}{m} - \dfrac{\rho A\, C_d\, v^2}{2m}} = \mathrm{d}t$$

Let

$$\alpha^2 = \frac{2(F - R)}{\rho A\, C_d}$$

and multiply both sides by $\dfrac{\rho A\, C_d}{2m}$:

$$\frac{\mathrm{d}v}{\alpha^2 - v^2} = \frac{\rho A\, C_d}{2m}\mathrm{d}t$$

$$\frac{1}{2\alpha}\left(\frac{1}{\alpha + v} + \frac{1}{\alpha - v}\right)\mathrm{d}v = \frac{\rho A\, C_d}{2m}\mathrm{d}t$$

This equation can now be integrated to give

$$\frac{1}{2\alpha}\left[ln\left(\frac{\alpha + v}{\alpha - v}\right)\right]_{v_2}^{v_1} = \frac{\rho A\, C_d}{2m}[t]_{t_2}^{t_1} \tag{5.8}$$

Case (a)

$$\alpha = \sqrt{\left(\frac{2(F - R)}{A\, C_d}\right)}$$

since F takes two constant values.

$$\alpha_1 = \sqrt{\left(\frac{2(3530-225)}{1.2 \times 2.25 \times 0.45}\right)} \qquad \alpha_2 = \sqrt{\left(\frac{2(1765-225)}{1.2 \times 2.25 \times 0.45}\right)}$$

$$= 73.76 \qquad\qquad = 50.35$$

and

$$\frac{\rho A\, C_d}{2m} = \frac{1.2 \times 2.25 \times 0.45}{2 \times 1000} = 0.6075 \times 10^{-3}$$

The acceleration time has to be evaluated in two stages:

(a) 0–13.89 m/s (50 km/h) in time t_a

(b) 13.89–27.78 m/s (100 km/h) in time t_b.

(i)
$$\frac{1}{2\alpha_1}\left[ln\left(\frac{\alpha_1 + v}{\alpha_1 - v}\right)\right]_0^{13.89} = \frac{\rho A\, C_d}{2m}[t]_0^{t_a}$$

$$t_a = \frac{1}{2 \times 73.76 \times 0.6075 \times 10^{-3}}\left[\left(ln\frac{73.76 + 13.89}{73.76 - 13.89}\right) - 0\right]$$

$$= 4.25 \text{ s}$$

(ii)
$$\frac{1}{2\alpha_2}\left[ln\left(\frac{\alpha_2 + v}{\alpha_2 - v}\right)\right]_{13.89}^{27.78} = \frac{\rho A\, C_d}{2m}[t]_0^{t_b}$$

$$t_b = \frac{1}{2 \times 50.35 \times 0.6075 \times 10^{-3}}\left[ln\left(\frac{50.35 + 27.78}{50.35 - 27.78}\right) - ln\left(\frac{50.35 + 13.89}{50.35 - 13.89}\right)\right]$$

$$= 11.04 \text{ s}$$

Total acceleration time $= t_a + t_b = 4.25 + 11.04 = 15.29$ s

Case (b)

$$C_d = 0.33$$

thus

$$\alpha_1 = 86.13$$

and

$$\alpha_2 = 58.80$$

$$\frac{\rho A\, C_d}{2m} = 0.4455 \times 10^{-3}$$

Again, the acceleration is evaluated in two stages:

$$t_a = \frac{1}{2 \times 86.13 \times 0.4455 \times 10^{-3}}\left[ln\left(\frac{86.13 + 13.89}{86.13 - 13.89}\right)\right]$$

$$= 4.24 \text{ s}$$

$$t_b = \frac{1}{2 \times 58.80 \times 0.4455 \times 10^{-3}}\left[ln\left(\frac{58.80 + 27.78}{58.80 - 27.78}\right) - ln\left(\frac{58.80 + 13.89}{58.80 - 13.89}\right)\right]$$

$$= 10.41$$

The total acceleration time is now 14.65 s, a reduction of 0.64 s or 4.2 per cent. As would be expected for the reasons given in the discussion of figure 5.12, the differences in acceleration times for 0–50 km/h are negligible.

In general, the available tractive force will also be a function of speed, and this will usually prevent the equation of motion being solved analytically. Instead a numerical approach will be needed, using the equations derived here. The acceleration times for a change of velocity have to be evaluated over a velocity increment for which the available tractive force (F) can be treated as a constant. The acceleration times for all the velocity increments can then be summed to give the total acceleration time.

Example 5.2

For the vehicle with parameters defined by figure 5.12, there is a residual tractive force (the difference between tractive force available from the engine and the total tractive resistance) of 550 N when travelling at 120 km/h. What is the maximum headwind into which the speed of 120 km/h can be maintained?

Solution

$$\text{Drag, } D = \tfrac{1}{2}\rho A\, C_d\, v^2, \text{ thus } D_{120+w} - D_{120} = 550 \text{ N}$$

where D_{120} is the drag at 120 km/h and D_{120+w} is the drag at 120 km/h with a headwind of speed w.

Thus

$$550 = \tfrac{1}{2}\rho A\, C_d \frac{(120 + w)^2 - 120^2}{(60^2 \times 10^{-3})^2}$$

$$550 = \frac{\tfrac{1}{2} \times 1.2 \times 2.25 \times 0.33}{(60 \times 60 \times 10^{-3})^2} \times (w^2 + 240w)$$

or

$$w^2 + 240w - 16 \times 10^3 = 0$$

of which the positive solution is $\underline{w = 54 \text{ km/h}}$.

Alternatively, at 120 km/h:

$$D_{120} = \rho A\ C_d\ v^2$$

$$= \tfrac{1}{2} \times \tfrac{1}{2} \times 2.25 \times 0.33 \left(\frac{120 \times 10^3}{60 \times 60}\right)$$

$$= 495\ \text{N}$$

Thus the total force available to overcome drag is $495 + 550 = 1045$ N.

$$D_{120+w} = \tfrac{1}{2} A\ C_d \left[\frac{(120 + w) \times 10^3}{60 \times 60}\right]^2$$

$$120 + w = \frac{60^2}{10^3} \times \sqrt{\left(\frac{1045}{\frac{1}{2} \times 1.2 \times 2.5 \times 0.33}\right)}$$

Taking the positive value again: $w = 54$ km/h.

5.8 Discussion points

(1) Identify the types of aerodynamic drag, and state their relative significance.

(2) What, apart from drag reduction, needs to be considered in the aerodynamic design of vehicles?

(3) State the principal techniques for reducing the aerodynamic drag of passenger cars.

(4) If the value of $A\ C_d$ for the vehicle discussed in figure 5.18 is reduced from 5.7 m^2 to 4.1 m^2, calculate the potential improvements in the steady-state fuel consumption for the laden and unladen cases, at speeds of 60 and 100 km/h. For the laden vehicle, what is the increase in maximum speed with 100 kW propulsive power?

(5) What are the methods for reducing the drag of trucks with separate tractors and trailers?

(6) A high-performance saloon car has a drag coefficient of 0.37, and a frontal area of 2.44 m^2. The effective mass of the vehicle and driver is 2000 kg, and the ambient conditions are an air temperature of 20°C and a pressure of 101.325 kN/m^2. In the speed range of 100–200 km/h, the rolling resistance is 340 N and the mean torque at the rear wheels is 1123 N m; the effective diameter of the tyres is 0.58 m.

Show that if the drag coefficient was reduced to zero, then there would be a 29 per cent reduction in the 100 to 200 km/h acceleration time.

6 Vehicle Design

6.1 Introduction

The aspects of vehicle design that are to be discussed in this chapter are those remaining that influence the fuel economy. The powertrain has been covered by chapters 2–4, and vehicle aerodynamics was treated in chapter 5; this chapter covers rolling resistance and vehicle mass.

The term 'rolling resistance' will include here the losses at the wheel bearings, the losses caused by rubbing brakes and, most significant of all, the losses associated with the tyres. Since under fixed conditions the tyre rolling resistance is almost directly dependent on the vehicle weight, any weight savings will lead to a direct saving in the constant-speed fuel economy. However, any weight reduction will have a more significant effect in reducing the fuel consumption during acceleration; this is, of course, much more difficult to quantify.

The reductions in weight lead to reductions in the size of the engine, transmission, suspension and other items in the running gear, all of which contribute to the vehicle weight. This self-feeding effect or weight compounding is subject, like almost everything else, to a law of diminishing returns; in other words, if a vehicle is made 10 per cent lighter it does not follow that the engine weight can also be reduced by 10 per cent. While the vehicle weight can be reduced, the payload, of course, cannot be reduced. This means that the percentage weight variation between a car being full and empty is becoming much greater, especially for small cars. If the vehicle handling is not to vary significantly, an active suspension system will be needed in which the suspension damping and stiffness characteristics can be varied according to the vehicle weight and ride.

Reductions in vehicle weight are rarely obtained by a radical change in technology. The weight savings are attributable to careful material selection (plastics of many types, ceramics, high-strength steel and thinner glass) and more thorough design. Computer Aided Design (CAD) and Computer Aided Manufacture (CAM) have made possible a much more integrated approach to vehicle design. In particular, finite element modelling (FEM) of the body and other components enables significant weight reductions to be obtained for a given design specification.

6.2 Rolling resistance

6.2.1 Brakes and bearings

The contributions of rolling resistance from the brake drag, wheel bearings and tyres will each be discussed in turn.

In the case of drum brakes, the brake shoes are withdrawn by springs after brake application. In contrast, disc brakes do not usually have any mechanism to withdraw the brake pads from contact with the disc after brake application. This leads to the brake pads rubbing the disc, and Hoffman and Beurmann (1979) quote the average brake drag torque as being about 3 N m. For a typical car with disc brakes on the front wheels, the brake drag torque contributes about 20 N to the rolling resistance–about 10 per cent of the total rolling resistance. Hoffman and Beurmann (1979) also describe a new design of brake caliper, in which the seal of the actuating piston causes the piston brake and pad to retract after the brakes have been used. This design is said to reduce the drag torque to about 0.7 N m.

Rolling element bearings are invariably used as wheel bearings, and they have inherently low friction values. However, to retain the lubricant in the bearing and to prevent the ingress of dirt into the bearing, seals are used. The usual type of seal has a lip that is in contact with the rotating element, and this is the most significant source of friction in the bearing assembly. The lip seal may account for up to about 10 per cent of the rolling resistance.

6.2.2 Tyres

Since the rolling resistance of a tyre is widely held to be a linear function of the tyre load, different tyres can be compared by using the tyre rolling resistance coefficient:

$$C_R = \frac{\text{tyre rolling resistance}}{\text{tyre load}}$$

Typically, the tyre rolling resistance coefficient lies between 1 and 1.5 per cent. Tyre rolling resistance is difficult to measure on roads, since the coarser the road surface, the greater the rolling resistance. New concrete and rolled asphalt have similar characteristics but polished concrete will reduce the rolling resistance by about 10 per cent, while gritted asphalt increases the rolling resistance by up to 10 per cent or more [DeRaad (1977)].

In order for any test results to be repeatable, the environmental conditions also have to be kept constant, and this is another reason for conducting tests in the laboratory. Since it is easier to react a tyre against a rotating drum than a continuous flat surface, most laboratory tests use a rotating drum. This is in contrast to the twin roll dynamometers that are used for emissions or fuel-consumption testing of vehicles; both types of rolling road are shown in figure 6.1.

Since the rolling resistance is going to depend on the degree of tyre deformation, the twin roll dynamometer would indicate the highest rolling resistance, as the degree of conformance with the tyre is least. The following correlation is widely used to interpret the readings from rotating drum tests–it is not applicable to twin roll tests:

$$R_{td} = R_{tf}\sqrt{(1 + D_t/D_d)} \tag{6.1}$$

where R_{td} = tyre rolling resistance measured on the drum
R_{tf} = tyre rolling resistance measured on the flat
D_t = effective rolling diameter of the tyre
D_d = drum diameter.

Before any further discussion of the factors that affect rolling resistance, it is as well to know the effect of any reductions in rolling resistance on the vehicle performance. For steady-state tests it is reasonable to assume that a given reduction in tractive force will lead to the same reduction in fuel consumption. For the car defined by figure 5.12 with a drag coefficient of 0.33, the effect of a 20 per cent reduction in rolling resistance is shown for a range of speeds in table 6.1.

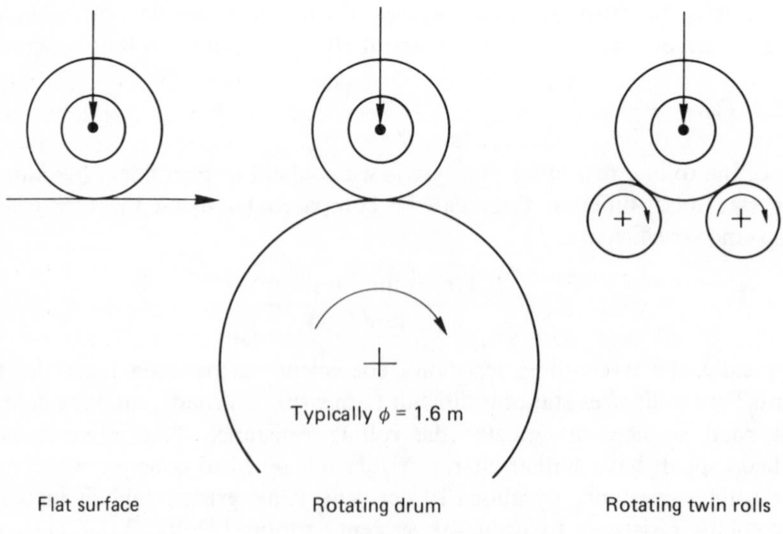

Figure 6.1 Methods for simulating road loads on tyres

Table 6.1 Automotive fuel consumption reduction at constant speed, for a 20 per cent reduction in the rolling resistance

	Vehicle speed (km/h)			
	40	80	120	160
Tractive resistance, baseline (N)	275	440	715	1100
Tractive resistance, 20% reduction in rolling resistance (N)	253	418	693	1078
Fuel consumption reduction (%)	8.0	5.0	3.1	2.0

Under normal driving conditions, the gains will be smaller because of the effects of speed variations. At high speeds the variations in speed will be small, so the steady-state estimate may be reasonable. At low speeds the variation in speed will be greater, and then the difference between the actual and steady-state fuel consumption gains will be greatest. Thus, where reductions in rolling resistance appear to offer the greatest gains, the effect of real (as opposed to steady-state) driving conditions will have the greatest effect in reducing the gains.

This indicates why the reductions in fuel consumption, for a given reduction in rolling resistance, should be independent of whether a highway or urban driving cycle is used. The results of Thompson and Torres (1977) imply that a 20 per cent reduction in rolling resistance would lead to a 4 per cent improvement in fuel economy.

Since the contribution of rolling resistance to the total tractive resistance is usually small, reducing the rolling resistance will have a negligible effect on the acceleration times. (It has already been shown that reducing the aerodynamic resistance has a negligible effect on vehicle acceleration, see example 5.1 in section 5.7, and the changes in rolling resistance will usually be smaller.)

The greatest potential for fuel-consumption improvements will be with heavy commercial vehicles for two reasons. Firstly, the rolling resistance will be of greater significance and, secondly, the maximum speeds will be comparatively low and the operating speeds will be more constant. For the laden truck defined by figure 5.18, the effect of a 20 per cent reduction in rolling resistance is shown in table 6.2. The calculation is the same as for table 6.1.

As before, the steady-state reductions in the fuel consumption at the highest speeds will be least affected by the reduction in the improvements attributed to the non-steady speed operation. For normal operation, the average truck

Table 6.2 Truck (mass, 32 500 kg) fuel-consumption improvements at constant speed for a 20 per cent reduction in rolling resistance

	Vehicle speed (km/h)		
	40	80	120
Tractive resistance, baseline (kN)	2.67	3.94	6.05
Tractive resistance, 20% reduction in rolling resistance (kN)	2.22	3.49	5.60
Fuel consumption reduction (%)	16.9	11.4	7.40

would obtain a 7 per cent improvement in fuel consumption for a 20 per cent reduction in rolling resistance [Luichini (1983)].

Now that the fuel economy benefits attributed to rolling resistance reductions have been established, it is appropriate to describe the factors that control the rolling resistance coefficient. The most significant difference is in the method of tyre construction–whether it is cross-ply or radial. The differences in tyre construction can be seen in figure 6.2. The cross-ply tyre has fabric reinforcements with textile chords making an angle of 20–30° with the circumferential centreline. In contrast, the radial tyre has textile chords that are at right angles to the circumferential centreline, with additional steel bracing under the tread.

The differences in rolling resistance between cross-ply and radial tyres are difficult to establish, since there is a wide variation between tyres of the same type produced by different manufacturers (±25 per cent), and furthermore there is also a significant variation between supposedly identical tyres [Thompson and Torres (1977)]. However, inspection of data from many sources suggests that radial tyres offer a 20–30 per cent reduction in rolling resistance when compared with cross-ply tyres.

The tyre shape is also important, since the larger the diameter of the tyre and the narrower its width, then the lower the value of the rolling resistance coefficient. Similarly, the smaller the tread depth, the smaller the rolling resistance coefficient. The rubber composition is also important, since this controls the hysteresis characteristics. As the tyre rotates it is continuously being loaded and unloaded, thus work will be dissipated by the hysteresis effects. The temperature of the tyre does not of itself appear to have much influence on the rolling resistance coefficient. However, since the volume of air in the tyre is almost constant, any rise in temperature raises the inflation pressure.

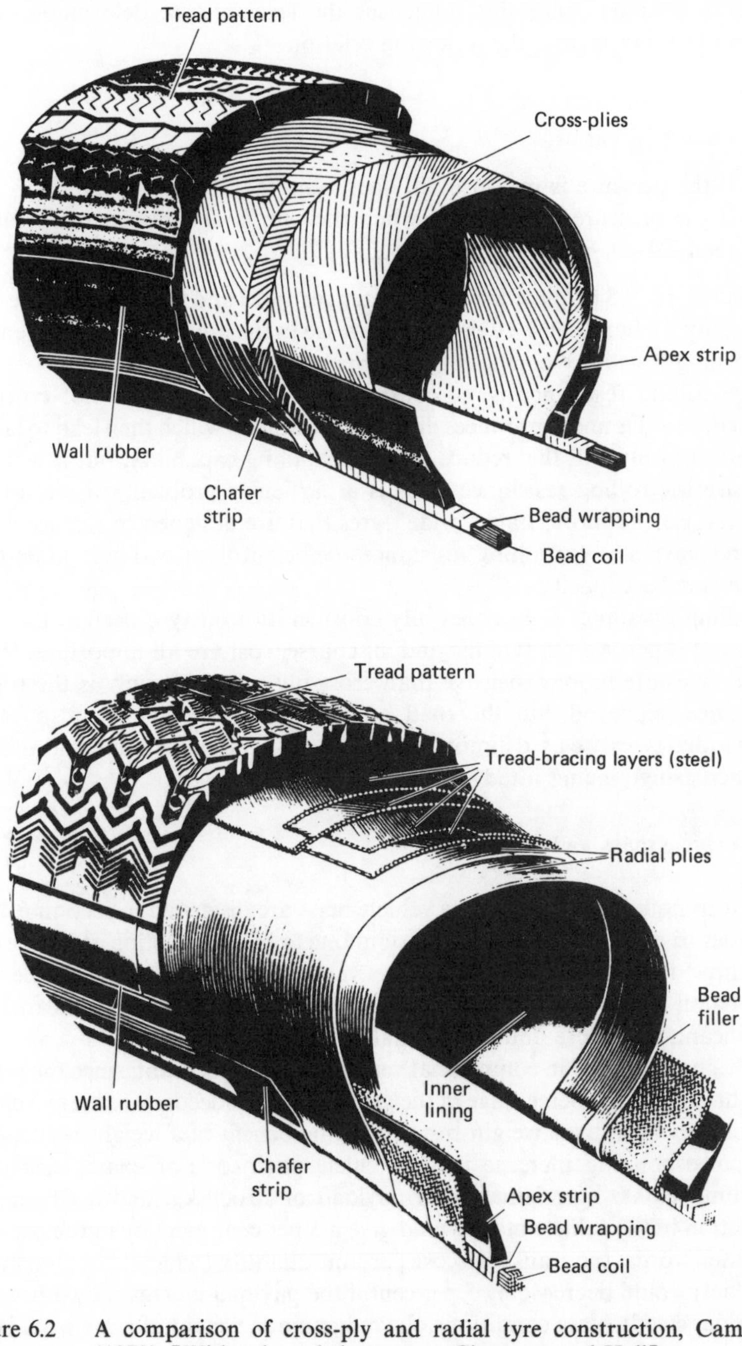

Figure 6.2 A comparison of cross-ply and radial tyre construction, Campbell (1978). [With acknowledgement to Chapman and Hall]

The rolling resistance coefficient (C_R) is quite sensitive to the inflation pressure (p, bar), since this influences the level of tyre deformation; and Klamp (1977) suggests the following relation:

$$C_R = 1/p^n \tag{6.2}$$

The value of n varies:

(a) If the pressure is measured in operation, then $0.6 < n < 0.9$.
(b) If the pressure is measured when the tyre is at ambient temperature, then $0.4 < n < 0.7$.

Thus if $n = 0.5$ (typical for a passenger car tyre) and the absolute pressure is initially 2.8 bar, then a 10 per cent increase in the inflation pressure reduces the rolling resistance coefficient by 4.7 per cent.

The rolling resistance is, of course, increased if tyres are not correctly aligned. Misalignment produces different slip angles, which then lead to lateral forces; not only will this reduce the road holding capability, but it will also increase the rolling resistance. This is a particular problem on multi-tyred vehicles. Furthermore, single 'wide' tyres that are designed to replace a pair of tyres have a lower rolling resistance coefficient than two individual tyres of the smaller capacity.

Rolling resistance is not the only consideration in tyre design: the road holding properties, the tyre life and, of course, cost are all important. While radial tyres are more expensive than cross-ply tyres, not only is the rolling resistance improved, but the road holding and durability are also better. Thus radial tyres are used almost universally on cars, and commercial vehicles are increasingly being fitted with radial tyres.

6.3 Vehicle mass and performance

The techniques for reducing the vehicle mass are discussed in section 6.4, but in order to evaluate the benefits attributable to mass reduction, it is necessary to know the trade-offs between the vehicle mass and performance. The discussion here will concentrate on cars, since the potential benefits are more significant, and more data are available.

Weight reduction in commercial vehicles is less significant, since the vehicle weight is a smaller percentage of the gross weight. Indeed, since many vehicles are limited by a gross weight restriction, the benefit of a weight reduction is the corresponding increase in the vehicle payload. For example, a truck weighing 10 000 kg might have a payload of 20 000 kg, and a 10 per cent reduction in the vehicle mass would give a 5 per cent increase in the payload. In other words, the transport cost per unit quantity (which is obviously not just fuel) would decrease by 5 per cent if the payload increases by 5 per cent.

Under steady-state conditions, the reduction in weight of a car would only improve the fuel consumption as a consequence of the reduction in rolling resistance. A 20 per cent weight reduction would reduce the rolling resistance by 20 per cent, and in section 6.2.2 it was suggested that this would give a 4 per cent improvement in fuel consumption. However, it is well known that constant-speed fuel-consumption figures are a poor indicator of actual fuel consumption; this is because of acceleration and hill climbing. Weight reduction will lead to a direct improvement of performance during hill climbing and acceleration, but the improvement may take the form of more rapid ascents and acceleration, rather than a fuel saving. By inspection of example 5.1 in section 5.7, it can be seen that a 20 per cent reduction in the effective mass would give at least a 20 per cent reduction in the acceleration time. This is discussed further and shown explicitly in the example in section 6.7.

One method of assessing the effect of weight reduction on fuel economy is to examine the data for a range of vehicles within a given class. This approach was used by Hanson (1979), who looked at the weighted EPA (Environmental Protection Agency) test cycle fuel consumption figures. The conclusion from this work was that reductions in fuel consumption were directly proportional to reductions in the effective mass. Thus a 20 per cent saving in mass would suggest a 20 per cent saving in fuel consumption, of which presumably 4 per cent or so could be attributed to the reduced rolling resistance.

The obvious drawback with this approach is that dissimilar vehicles are being compared. In a given class, the lighter vehicles are also likely to have better matched powertrains and better aerodynamic shapes. Thus the improvements in fuel consumption are not solely attributable to reductions in mass.

Results from a specific car are quoted by Porter (1979). For this medium-sized US saloon, a 20 per cent reduction in mass led to a 15 per cent improvement in the weighted EPA fuel consumption. Again, 4 per cent of the improvement can probably be attributed to the reduction in rolling resistance. Care should also be taken when applying these results to European cars, since US cars are in general larger and much heavier. Dorgham (1982) suggests that for a European car, a 20 per cent weight saving would only give an 8 per cent improvement in fuel consumption.

Finally, it must be emphasised that the fuel savings obtained in practice will depend very strongly on the vehicle usage. The limiting lower case is where a car is mostly driven on the flat at constant speed (an approximation to motorway driving), and under these conditions the fuel savings will tend to those attributable to the reduction in rolling resistance.

6.4 Vehicle design techniques

To produce a vehicle with a low mass, a rigorous approach is needed in vehicle design. The savings in mass can come from improved designs with existing materials or the use of alternative new materials. The reducing cost and increasing power of computers has had a major impact on vehicle design, and the most significant technique for vehicle design must be finite element modelling (FEM). In the finite element method, mechanical components are modelled by a mesh that consists of either triangular or rectangular elements, an example is the suspension arm shown in figure 6.3. This component is pressed from sheet material, and the assumption is that the stress is constant through the thickness of the material. In an object such as a cylinder block a three-dimensional mesh is needed, in which case tetrahedrons are most commonly used.

The finite element method determines the relationship between stress and strain for each element in such a way that the external constraints are met, and that there is compatibility between adjacent elements. The finite element model can thus be used to determine the stress levels in components for a given loading, and the deflections of different parts of the component. The use of the model can be extended to predict vibration modes and frequencies, also the effects of impact, and to predict heat flow patterns.

The advantage of the finite element method is that it enables the stress to be determined at all points in a component, and this leads to more efficient design. Since the shape of a component can be changed quite readily, it is also much easier to analyse a range of designs than by building and testing, consequently this leads to a more efficient design. The finite element method enables designs to be analysed more accurately and comprehensively, but this is of no advantage if the various loads and permissible deflections are not equally well defined. The static and dynamic loads need to be established from all possible loadings. For example, in a suspension arm, there may be load inputs from the suspension, powertrain and steering.

The deflections of a component or system can be determined, but it is much more difficult to define what deflections will be acceptable. For example, the suspension or its mountings must not deflect such that the steering geometry is affected by cornering loads. Similarly, the vehicle body must be sufficiently stiff, such that if a front wheel is jacked up with the adjacent door open, then the windscreen must not pop out. These problems have arisen because the acceptable stiffnesses and deflections were not known.

Computer technology also enables the whole design process to be conducted more expediently. In general, the body shape is developed using a full-size clay model, and this can be measured automatically. The data are recorded by a computer, and surface-fitting routines can be used to remove imperfections in the model and any asymmetry. Since the data that define the body shape are stored, they are also available for other uses–for example, aerodynamic modelling (see section 5.5), producing scale models for wind tunnel testing, generating meshes for finite element analysis, and 'driving' numerically controlled machine tools that produce the dies used in pressing

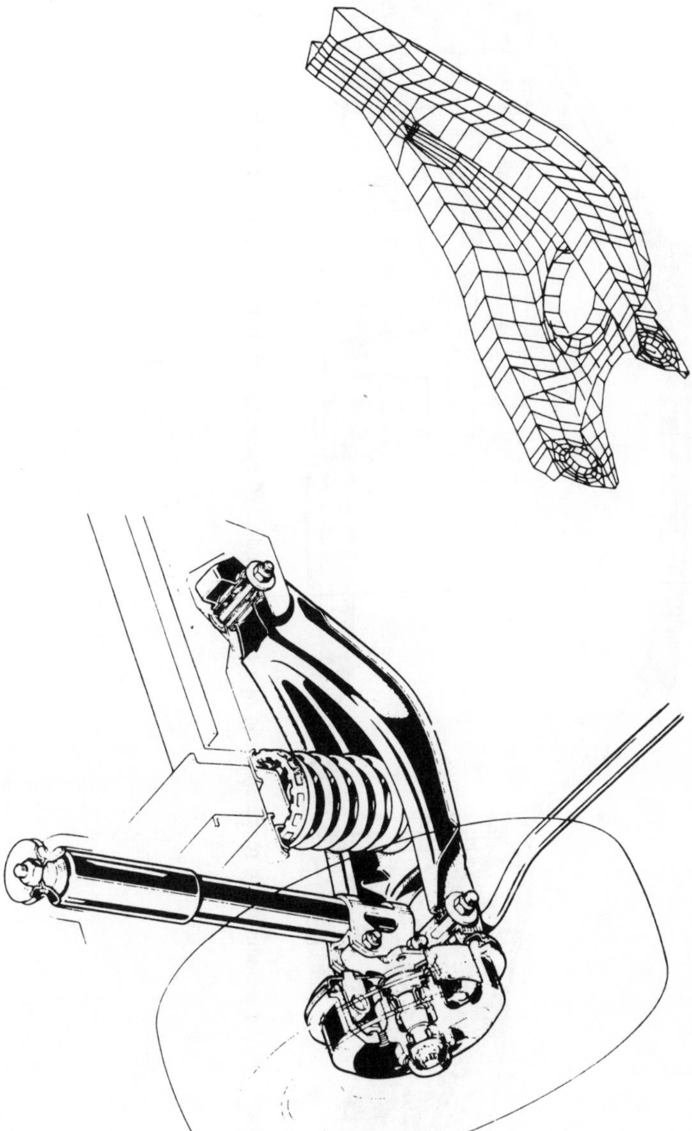

Figure 6.3 Example of a component (a suspension arm), with its finite element mesh representation. [Reproduced by permission of Ford Motor Co. Ltd]

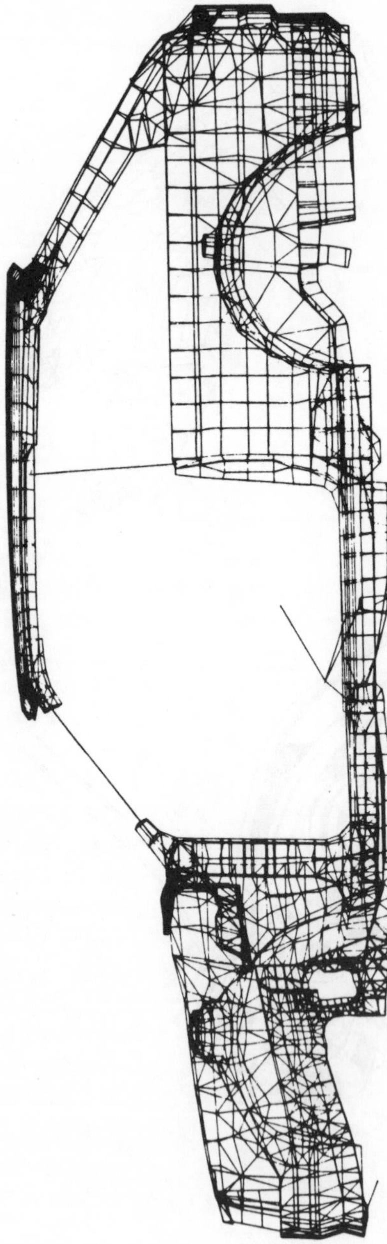

Figure 6.4 Finite element mesh representation of a car body side. [Reproduced by permission of Ford Motor Co. Ltd]

the body components. The finite element mesh for a car body side is shown in figure 6.4

6.5 Materials selection and use

6.5.1 Introduction

Designing for a low mass is not the only consideration in vehicle design. Very often it is the manufacturing considerations that dictate the materials used in vehicle construction. Mass-produced cars are normally made from pressed steel components that are welded together, and this has needed a heavy investment in press and welding equipment. Since the total investment is very high, the equipment tends to be updated on a continuous basis; for example, robot installations are now widely used for welding the body components together. Suppose a suitable reinforced plastic material was available for making car bodies, with a much lower material cost. No doubt such body components would be held together by adhesives, in which case the robots would be readily adapted for the purpose. However, the heavy investment in the sheet metal presses would be lost, and a substantial new investment would be needed in plastic moulding equipment.

These arguments apply to high-volume production, where plastic panels would need to be produced at the same rates at which presses can stamp steel body panels. For low-volume car production (say one per day as opposed to a thousand per day) the position is quite different. Glass fibre reinforced plastic (grp) panels can be laid-up by hand, in hand-made moulds. The process involves coating the mould with a gel coat (that is ultimately the outer surface and can thus be self-coloured), and then adding successive layers of glass fibre mat or cloth which are then impregnated with a polyester resin *in situ*. This is a highly labour-intensive process with a low output rate, but the capital investment is very low; it is a semi-skilled process, and the designs can be readily changed.

The discussion of materials selection in this section is meant as an introduction, with illustrations of mass savings from a few specific examples. A useful treatment of materials selection is given by Ashby and Jones (1980), who consider material properties, energy content and cost. The criteria for selecting materials for different applications are also developed in a rigorous way; for example, the relative mass and cost are presented for different material beams of a given stiffness.

6.5.2 Metallic materials

Mild steel has been used almost exclusively for the bodies and other structural components of mass-produced vehicles. It is a cheap material, for which the manufacturing processes have become highly developed. Mild steel can be welded quite easily, and it is also suitable for pressing panels with deep drawn shapes, since after elastic yield it can undergo extensive plastic deformation before fracture. Furthermore, as mild steel has a comparatively low yield strength, the tendency for the shape to revert or spring back after the pressing operation is minimal. However, the drawbacks with mild steel are its comparatively high ratio of weight to strength, and its tendency to rust. The corrosion of mild steel can be minimised by many processes–for example, zinc coatings, cathodic electrocoats and wax coatings.

Apart from plastics, the materials that need to be considered for car bodies and structural components are aluminium alloys and high-strength steel. High-strength steel has the same corrosion problems as mild steel, while aluminium alloys have good corrosion resistance, especially if they are anodised. However, the greater cost and the manufacturing problems associated with aluminium alloys dictate against their use. Aluminium alloys are difficult to weld, and the shapes that can be pressed in a single operation from mild steel would require several operations if manufactured from an aluminium alloy. Some of the main material properties are summarised in table 6.3.

For a component of a given strength, an aluminium component would be the lightest, but it would also be the most expensive. Even for an unstressed body panel of the same dimensions, the aluminium panel would be the most expensive (that is, it has the greater cost per unit volume). Furthermore, such an unstressed panel would lend itself to being made from plastic.

As mentioned before, it is often a component's stiffness that is as important as its strength, and Young's modulus has been tabulated here since it gives the relation between load and deflection. Firstly, it should be noted that alloying the steel has not changed its Young's modulus. Secondly the Young's modulus for aluminium alloys is one-third that of steel. In other words, if three identical components were made from mild steel, high-strength steel

Table 6.3 Summary of material properties

Materials	Yield strength (MN/m^2)	Young's modulus (kN/mm2)	Approximte relative cost (cost per kg)	Density (kg/m^3)
Mild steel	180	210	100	7800
High-strength steel	280	210	110	7800
Aluminium alloy	170	70	400	2700
Cast iron	–	115	50	7270

and alumium alloy, for a given stress the deflections of the two steel components would be the same, and the deflection of the alloy component would be three times greater. For these reasons, aluminium alloys are not attractive for structural components or bodywork.

In contrast, high-strength steel is attractive for certain applications. Body panels cannot be reduced in thickness for two reasons: firstly corrosion, and secondly the stiffness needs to be maintained. However, high-strength steel is attractive for structural components such as engine mounts, subframes and commercial vehicle chassis (figure 6.5). The high-strength steel chassis has a weight saving of 100 kg, or about 33 per cent when compared with a design made from mild steel [Dorgham (1982)]. The Ford Sierra has some 35 kg of high-strength steel, and where this material is used it gives a 15–20 per cent weight saving.

When a component is being redesigned for high-strength steel, it is obviously not sufficient just to specify thinner materials with the original design. While such a component would be strong enough, the stiffness would no longer be adequate; a high strength steel component will require deeper sections. Current steel body construction techniques are described comprehensively by Jacobson (1984).

The position with cast components is somewhat different. Traditionally, the engine and transmission casings have been made from iron castings. Direct substitution of aluminium alloy is not possible, since while the strength would be comparable the stiffness would be lower. However, the use of finite element design methods enables components to be redesigned, with carefully

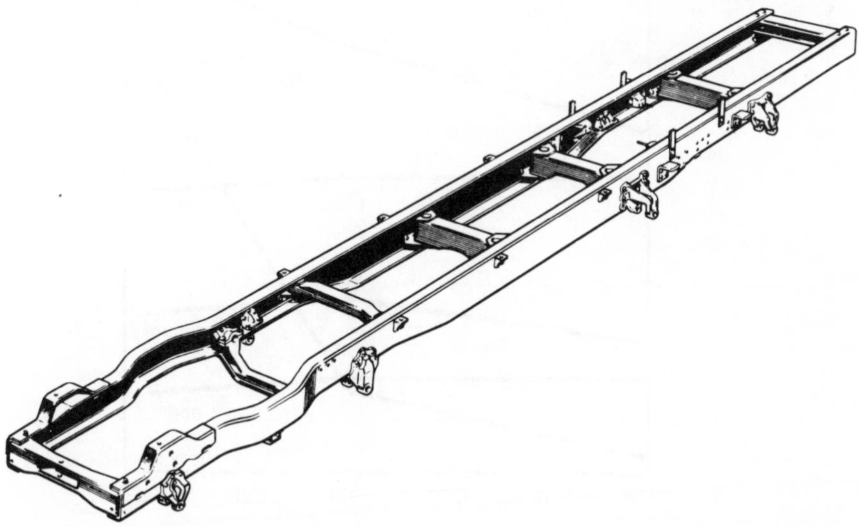

Figure 6.5 Example of high-strength steel application, the Ford Cargo truck chassis, Dorgham (1982)

placed webs and gussets to maintain the stiffness. The attraction of alumium alloy castings is the significant weight saving, perhaps as much as 100 kg on a large car.

Table 6.3 shows that the difference in stiffness between cast iron and aluminium is not significant, but the price difference is a factor of 8. However, this needs to be offset against reduced foundry costs associated with lower casting temperatures (about 1000°C compared with about 1500°C), and the potential for recovery of alumium alloy from scrap cars.

6.5.3 Non-metallic materials

The recent trends in car material usage are shown in figure 6.6; this shows that the weight of plastics in a car is increasing significantly, and that the weight of glass is approximately constant. The current trends in vehicle design are for an increase in the glazed area, but since there has been a reduction in the glass thickness the overall weight is approximately constant. A medium-sized hatchback car will have a glazed area of about 3 m^2, with a thickness of typically 3 mm. If the glass thickness had not been reduced from earlier values of 5 mm, then there would be a weight penalty of 15 kg.

Not only has the glass thickness been reduced, but the method of locating

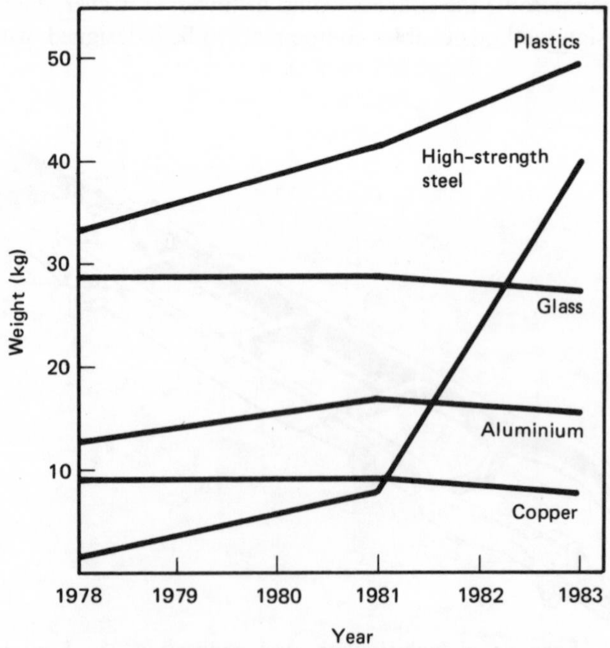

Figure 6.6 Average materials usage in cars, Dorgham (1982)

the fixed screens has also been changed; they are now surface mounted, as shown in figure 6.7. The conventional system used a rubber seal that was placed in the vehicle, and then the glass would be eased into the rubber. Finally, the glazing bead would be fed into the rubber to secure the glass. The surface-mounting system uses an adhesive to attach the glass to the vehicle structure, with the sealing strips performing a decorative and functional role. The surface-glazing system has several advantages: a better aerodynamic performance, the glass can contribute to the stiffness, and the adhesive and glass can both be applied by robots. Webb (1984) reports that direct glazing can increase the torsional stiffness of an unglazed four-door saloon car by as much as 75 per cent, and this offers scope for materials savings in the body.

The use of plastics for interior trim is well established, and new applications are being found for components associated with the running gear; examples include reservoirs, air cleaner housings, belt covers, many components that were previously die cast, fans, impellers, lightly loaded mountings and end tanks for radiators. Plastics are also finding wider use in exterior applications; the most common examples are bumpers, trim items, grills and to a lesser extent body panels.

Plastic bumpers are now a popular solution, with a useful weight saving that is quantified in section 7.2.3; the options include a single self-coloured bumper for the whole model colour range, matching self-coloured materials, and materials that will accept a paint finish. In selecting suitable materials, it is not just the mechanical properties of resistance to impact or deformation and energy absorption that need to be considered. Due account needs to be taken of the manufacturing processes available, the required cycle time, and whether or not the material has to withstand temperatures of up to 180°C in the paint baking ovens. Different bumper materials are listed in table 6.4 with their attributes.

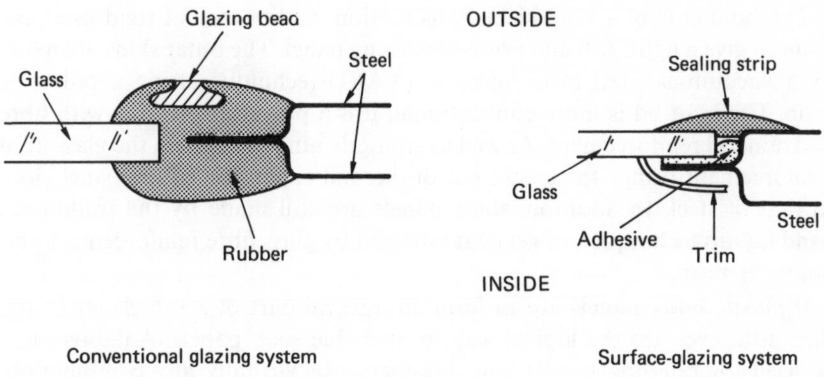

Figure 6.7 Comparison of glazing systems

Table 6.4 Different plastics used for car bumpers

Material	Atrributes
Sheet moulding compound	Cheap
Polypropylene	Comparatively cheap with good shape recovery
Polycarbonate	Twice the cost of polypropylene but self-finished with fast cycle times
Polyester	Expensive but able to withstand high temperatures

An area that still offers significant potential is the use of plastic body panels as part of the structure. However, the current approach is to have a steel inner body or space frame that acts as a chassis, with attachments for the powertrain and suspension, and mounting points for the body panels. Current examples of this approach are the Citroen BX (which also uses some steel panels), the Pontiac Fiero and the Reliant Scimitar SS1.

The construction of the Reliant SS1 is shown in figure 6.8, and the computer-aided design of the chassis is described by Fothergill *et al.* (1984). The body panels are manufactured from a range of materials according to the application, and since the panels are not stressed, the replacement of any damaged panel is simple. The front and back panels and wings are made from reinforced reaction injection moulded (RRIM) polyurethane. This process consists of a two-part polyurethane being pre-mixed with chopped glass fibre reinforcement. After being forced into a mould, the constituents react, expand and cure. These panels have good impact and damage resistance since they are semi-flexible.

The bonnet is of a sandwhich construction, with a core of rigid urethane foam to give a light, stiff and noise-absorbing panel. The outer skins are made by a vacuum-assisted resin injection (VARI) technique, using a polyester resin. The boot lid is more conventional; it is a polyester pressing with fibre and mineral reinforcement. As well as strengthening the panels, the glass fibre reinforcement brings the coefficient of thermal expansion of the panel close to that of steel. In addition, some panels are still made by the traditional hand lay-up techniques, of gel coat followed by glass fibre reinforcement and polyester resin.

If plastic body panels are to form an integral part of a vehicle structure, then adhesives are the logical way to assemble such panels. Adhesives can be used for bonding metals and non-metals in virtually any combination. Furthermore, adhesives can be used in conjunction with spot welding to

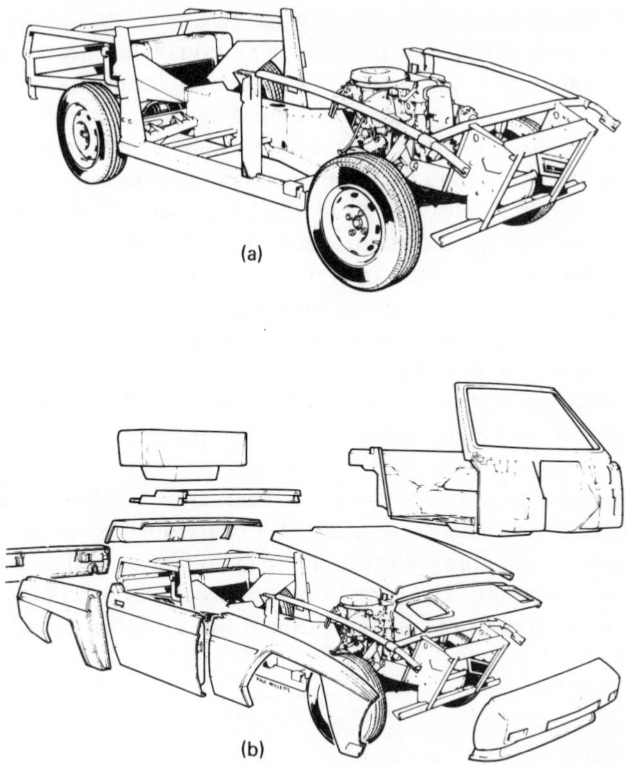

Figure 6.8 Construction of the Reliant Scimitar SS1: (a) space frame, (b) body panels. [Reprinted by permission of the Council of the Institution of Mechanical Engineers]

increase the strength and impact resistance of spot-welded panels and structures. The presence of the adhesive will also inhibit the corrosion in spot-welded seams. Another advantage of adhesives over welding is that there will be no residual stresses.

The selection of a suitable adhesive is, of course, essential. The adhesive needs to form a good bond with the materials that it joins, resistance to thermal and mechanical shock, and the ability to accept strains without brittle failure. The different types of adhesive and their characteristics are discussed by Lees (1985), with particular reference to adhesive selection.

Very important classes of adhesive are the toughened epoxy and toughened acrylic adhesives. High-strength adhesives are usually also brittle and hard, and any subsequent distortion of the bonded material leads to strong cleavage and peel forces that break the joints. By having a dispersed rubbery phase within the harder load-bearing adhesive matrix, the crack propagation is inhibited and the adhesive is toughened.

Mention has already been made of glass fibre reinforced plastic (grp) in the context of body panels, in which the orientation of the fibre reinforcement is essentially random. A much more demanding application of composites is in the suspension system, which is described by de Goncourt and Sayers (1985). One approach is to use a glass fibre reinforced composite to replace a metallic spring. A more radical approach is to use the new component to fulfil several additional roles, namely to take lateral loads, to act as an anti-roll bar and to take braking and acceleration loads. A composite suspension is shown in figure 6.9, alongside a conventional MacPherson strut suspension for comparison. The composite spring has continuous axial fibre reinforcement, and the cross-sectional area of the spring is constant. The width of the spring is varied in order to control the spring and roll characteristics. The glass fibres are embedded in an epoxy matrix that represents 30 per cent by weight of the composite, and a compression moulding process is used to eliminate voids.

The composite suspension described here gives a mass reduction of 67 per cent or about 10 kg, and since this is a reduction in the unsprung weight it leads to an improvement in the road holding as well as a reduction in the fuel consumption. The composite suspension also offers a significant reduction in the number of components, and this should also lead to a cost saving.

6.6 Conclusions

For good fuel economy and performance, the rolling resistance of a vehicle needs to be minimised. Since the rolling resistance is essentially proportional to the vehicle weight, then weight reduction leads to a reduction in rolling resistance and an improvement in vehicle acceleration. Any weight reduction also has a compounding effect, since a lighter vehicle has a lighter suspension, engine, transmission and braking system. However, variations in payload will have a greater effect on a lighter vehicle in terms of the handling acceleration and fuel economy.

The rolling resistance of a vehicle can also be reduced by reducing the rolling resistance coefficient of the tyres. A very significant development has been the introduction of steel-braced radial tyres, since these have a rolling resistance that is some 20–30 per cent lower than the rolling resistance of cross-ply tyres. The other parameters that affect the tyre rolling resistance are the inflation pressure, tread depth and the ratio of the diameter to width. Narrow tyres with a small tread depth and high inflation pressure have the lowest rolling resistance coefficient.

Reductions in the rolling resistance will have most effect on the fuel consumption at low vehicle speeds, since under these conditions the aerodynamic resistance is not yet significant. However, it is at the low speeds

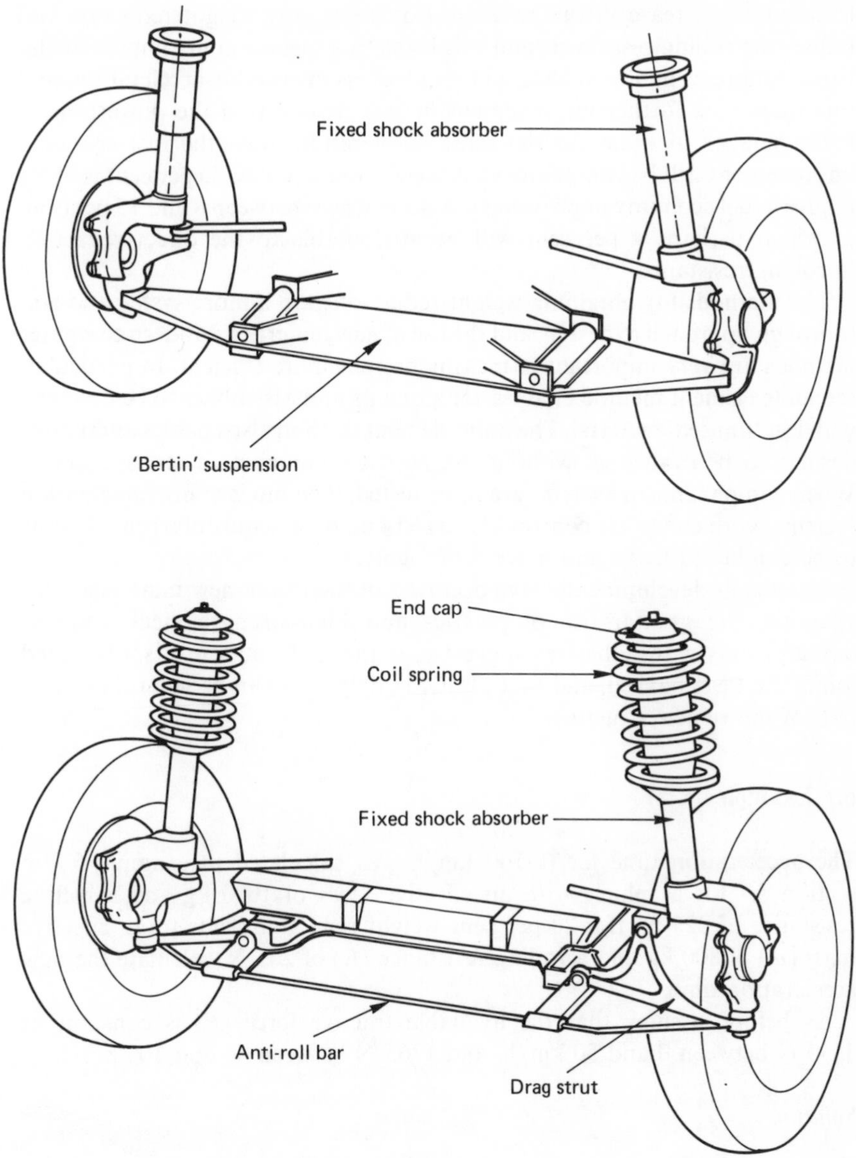

Figure 6.9 'Bertin' composite suspension

that more time is spent accelerating and braking. Thus, for a range of driving patterns, a 20 per cent reduction in rolling resistance would give about a 7 per cent improvement in truck fuel economy, and a 4 per cent improvement in car fuel economy.

For a truck, the reduction in vehicle weight is most significant when it leads to an increase in the payload. For a car, any weight reduction will reduce the rolling resistance, and this leads to a fuel economy improvement. Furthermore, a lighter vehicle will require less energy for acceleration, and this leads to a further improvement in fuel economy. If the powertrain is made smaller to maintain the same acceleration, then the fuel economy improvement will be compounded. A weight reduction of 20 per cent is likely to give a fuel economy improvement of somewhere between 8 and 15 per cent, of which perhaps 4 per cent will be attributable to the direct reduction in rolling resistance.

The methods for obtaining weight reductions are a more systematic and thorough approach to design, and the use of new materials. Modern computer methods are very important in making designs more efficient. In particular, the finite element method enables components and assemblies to be designed with less 'unused' material. The finite element method also enables alternative designs to be evaluated, without the need to manufacture test specimens. When components or systems are to be tested, then modern instrumentation systems (with computer control of the test and data acquisition) enable tests to be conducted faster and more thoroughly.

Significant developments have occurred in the use of new materials – for example, the greater use of plastics and high-strength steel. Current developments with adhesives suggest that the performance of spot-welded joints can be enhanced, and that plastic body panels can be incorporated as part of the vehicle structure.

6.7 Example

The acceleration time for 0–100 km/h was calculated in example 5.1 in section 5.7 for a vehicle with an effective mass of 1000 kg and a rolling resistance of 225 N. If a 20 per cent weight reduction leads to an effective mass (m) of 800 kg and a rolling resistance (R) of 205 N, calculate the new acceleration time.

As before, assume that the available tractive force (F) is constant at 3530 N between 0 and 50 km/h, and 1765 N between 50 and 100 km/h.

Solution

From equation 5.8:

$$\frac{1}{2\alpha}\left[ln\left(\frac{\alpha+v}{\alpha-v}\right)\right]_{v_2}^{v_1} = \frac{\rho A\ C_d}{2m}[t]_{t_2}^{t_1} \tag{5.8}$$

where

$$\alpha = \sqrt{\left(\frac{2(F-R)}{A\ C_d}\right)}$$

If the rolling resistance had been unchanged, then the left-hand side of equation 5.8 would be the same as in the present example, and a given percentage reduction in the mass would give the same percentage reduction in the acceleration time. However, since the rolling resistance has also been reduced, then the left-hand side of equation 5.8 has to be re-evaluated, and there will be another contribution to reducing the acceleration time.

$$\alpha_1 = \frac{2(3530 - 205)}{1.2 \times 2.25 \times 0.45} \qquad \alpha_2 = \frac{2(1765 - 205)}{1.2 \times 2.25 \times 0.45}$$

$$= 7.398 \qquad\qquad = 50.67$$

and

$$\frac{\rho A\ C_d}{2m} = \frac{1.2 \times 2.25 \times 0.45}{2 \times 800} = 0.759 \times 10^{-3}$$

Again, the acceleration time has to be evaluated in two stages:

(i) 0–13.89 m/s (50 km/h) in time t_a

(ii) 13.89–27.78 m/s (100 km/h) in time t_b.

(i) Rearranging equation 5.8 and substituting gives

$$t_a = \frac{1}{2 \times 73.98 \times 0.759 \times 10^{-3}}\left[ln\left(\frac{73.98 + 13.89}{73.98 - 13.89}\right) - ln\left(\frac{73.98}{73.98}\right)\right]$$

$$= 3.38 \text{ s}$$

(ii) Similarly for 13.89–27.78 m/s:

$$t_b = \frac{1}{2 \times 50.67 \times 0.759 \times 10^{-3}}\left[ln\left(\frac{50.67 + 27.78}{50.67 - 27.78}\right) - ln\left(\frac{50.67 + 13.89}{50.67 - 13.89}\right)\right]$$

$$= 8.69 \text{ s}$$

$$\text{Total time } (t_a + t_b) = 3.38 + 8.69 = 12.07 \text{ s}$$

Compared with example 5.1 of section 5.7, this is a percentage reduction of

$$\left(\frac{15.29 - 12.07}{15.29} \times 100\right) \text{per cent} = 21.06 \text{ per cent}$$

6.8 Discussion points

(1) How does a reduction in the vehicle mass improve the fuel economy of (a) trucks, (b) cars?

(2) What are the factors that influence the rolling resistance coefficients of tyres?

(3) How are reductions in vehicle mass achieved?

7 Case Studies and Conclusions

7.1 Introduction

The two case studies that are presented here have been chosen to illustrate the techniques that have been developed in this book. The first case study is based on the Rover 800, and the second case study examines the Ford Transit van. Both of these vehicle ranges have been designed for good fuel economy, and have achieved sales on a world-wide basis. In each case a brief description is given of the whole model range, before concentrating on a particular example of each.

The Rover 820i, which has been selected for a detailed study here, has a high efficiency spark ignition engine driving through a four-speed automatic gearbox. As a contrast, the Ford Transit van discussed here has a high-speed direct injection diesel engine, driving through a five-speed manual gearbox. Consequently, these two vehicles illustrate many of the powertrain options that have been discussed in chapters 2–4. Both vehicle ranges have also been the subject of extensive aerodynamic and structural design work. In each vehicle there are examples of modern design techniques and the use of high performance materials to give weight savings.

This chapter ends with a summary of the whole book; in other words, the way fuel economy is affected by the factors that are controlled in vehicle design.

7.2 The Rover 800

7.2.1 Introduction

The Rover 800 is shown in figure 7.1; it is a four-door front wheel drive saloon car, with independent front and rear suspension. This car was designed jointly by Austin Rover and Honda, as a result of an agreement signed in 1981. The Honda version (known as the Legend) and the Rover 800 share

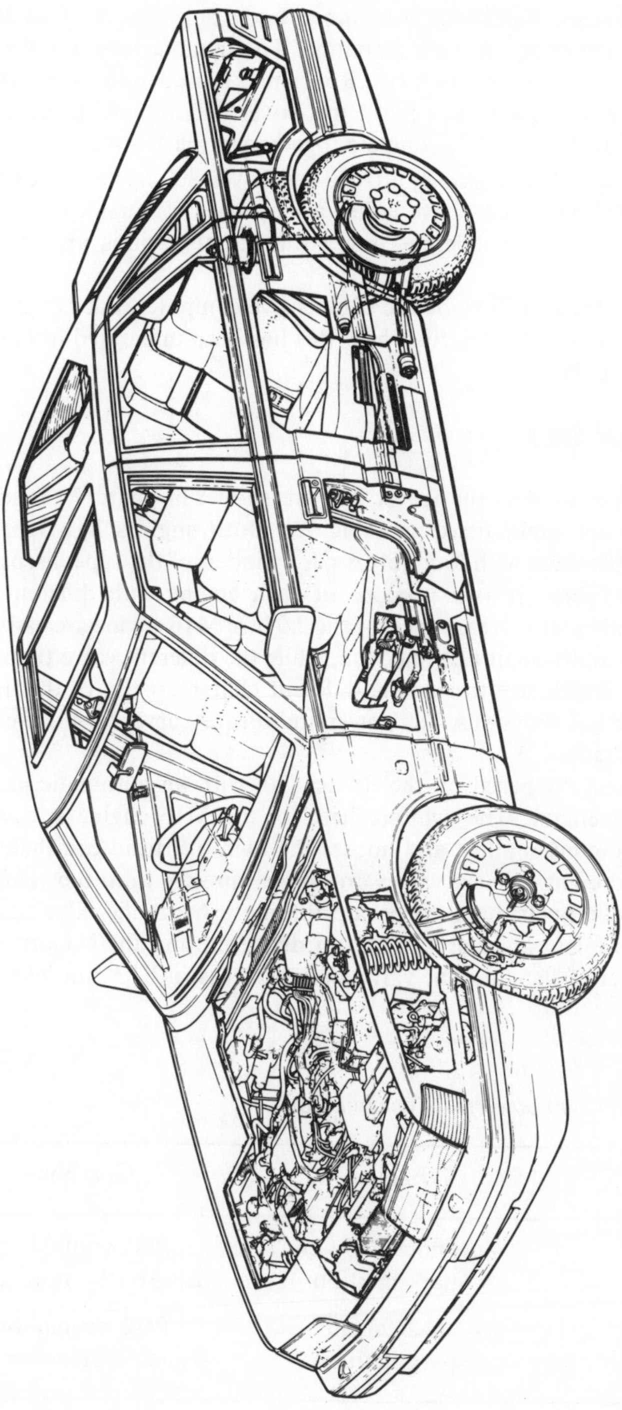

Figure 7.1 The Rover 800. [Copyright of Austin Rover Group Limited and reproduced with their kind permission]

the same floorpan, bulkheads and inner body sides, as well as having many mechanical components in common. The manufacture has also been undertaken on a collaborative basis. Rover make the Honda Legends that are sold in Europe, and Honda make the Rover 800 for sale in the Eastern Hemisphere. The two models compete together in the US market.

In addition to different levels of trim, the Rover 800 has a choice of two engines, each of which can have either a five-speed manual, or four-speed automatic gearbox; the different powertrain combinations are shown in table 7.1.

The Rover 800 has also made use of extensive computer-aided engineering (CAE), and section 7.2.3 describes how this has been integrated from design through to manufacture.

7.2.2 The Rover 800 powertrain

The two engines used in the Rover 800 are shown in figure 7.2, and their specifications are summarised in table 7.2. Both engines have pent-roof combustion chambers with four valves per cylinder, of the type discussed in section 2.4.4. There are two versions of each engine with different valve timings. The different valve timings for the 2.0 litre M16 engine are associated with single or multi-point fuel injection, while the different valve timings on the C25A V6 engine are to suit the different characteristics of the manual and automatic transmissions. The power and torque curves for these engines are shown in figure 7.3.

This case study is based on the Rover 820i with an automatic gearbox, and the M16I engine. The cylinder head of the M16 engine is cast in a low silicon aluminium alloy, and houses the twin overhead camshafts. The eight inlet and eight exhaust valves are driven directly from the camshafts, through bucket tappets that incorporate automatic hydraulic valve clearance adjusters. The inlet valves are 29 mm in diameter, and the exhaust valves have a diameter of 26.5 mm; the valve lift in each case is 8.83 mm. The valves

Table 7.1 Rover 800 powertrain combinations

Model	Engine	Gearbox
820E/SE 820i/Si	4-cylinder, 1994 cm^3 spark ignition	PG1 manual or ZF 4HP14 automatic
825i/Sterling	V6, 2494 cm^3 spark ignition	PG2 manual or Honda 4AT automatic

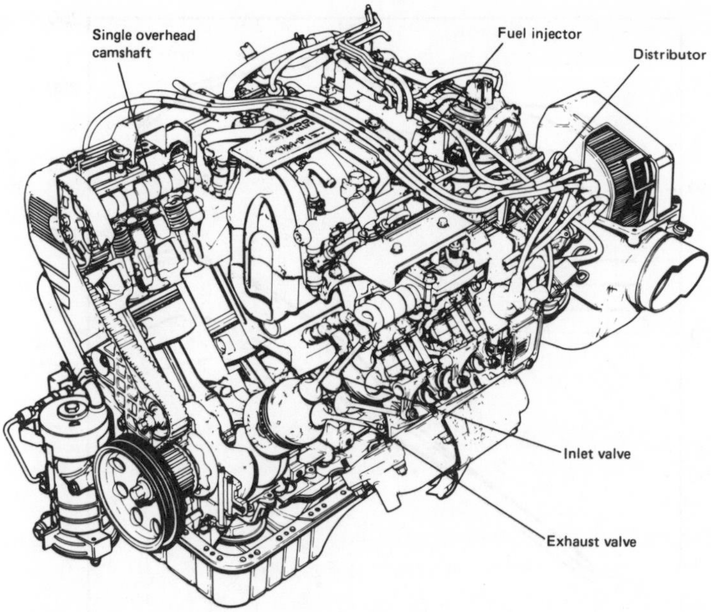

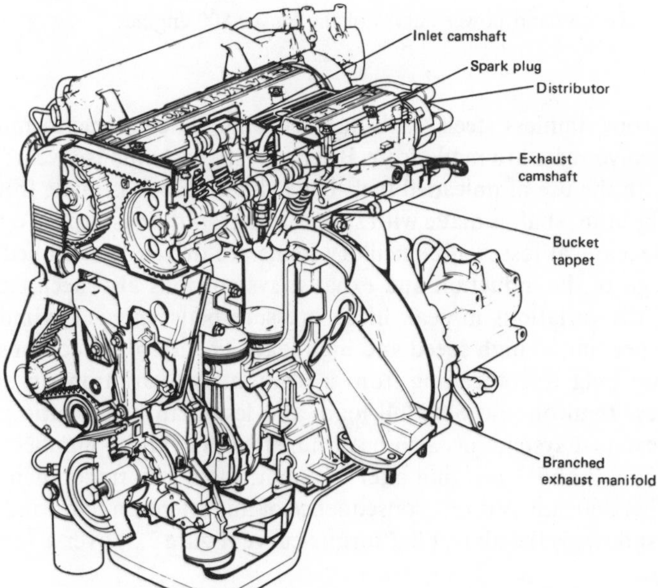

Figure 7.2 Engines used in the Rover 800. [Copyright of Austin Rover Group Limited and reproduced with their kind permission]

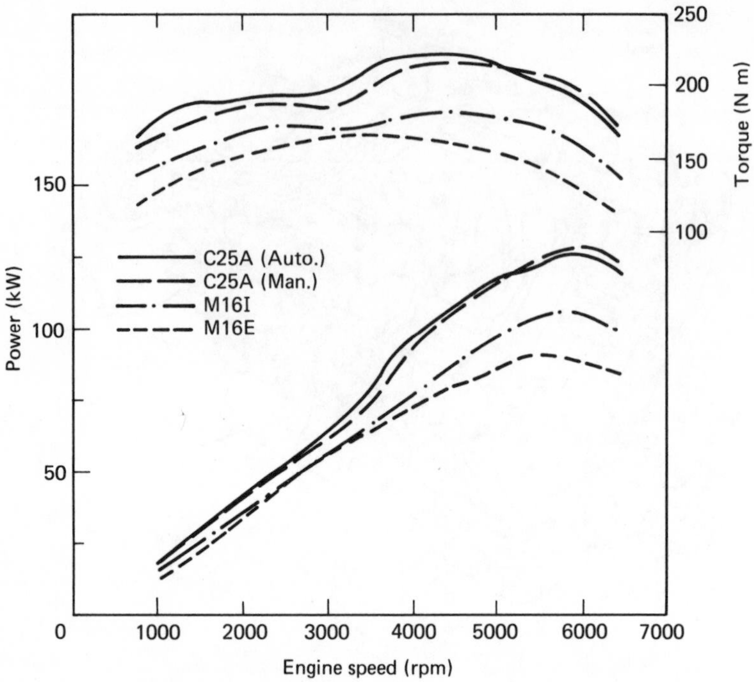

Figure 7.3 Torque and power curves of the Rover 800 engines

are made from stainless steels, with hard chrome-plated stems to minimise wear. The valve seats are made from Brico inserts, to minimise wear; this is essential with the use of unleaded fuel. The cylinder block is made from cast iron, and the crankshaft is made with spheroidal graphite cast iron. To reduce external mechanical losses, the ancillary items are driven by a ribbed belt.

The design of the induction and exhaust systems has also been carefully optimised; the variations in peak firing pressure between each cylinder are less than 5 per cent at high speed and high load. The M16I induction system has 292 mm long tracts leading from a plenum; the variations in air/fuel ratio are less than one ratio at full load, and less than half a ratio at part load. The exhaust system has also been tuned, with cylinder numbers 1 and 4 merging and 2 and 3 merging after 340 mm; these two tracts then merge after a further 860 mm. A direct consequence of this attention to the induction and exhaust flows is the almost flat torque curve (figure 7.3) over a very wide speed range.

The combustion chamber in the M16 engine is of the pent-roof type (figure 2.14) with four valves per cylinder and a central spark plug. The orientation of the inlet ports relative to the inlet valves is critical, since it influences the trade-offs between emissions, efficiency and power output. The

Table 7.2 Rover 800 engine specifications

Engine type	M16E	M16I	C25A (manual)	C25A (automatic)
Cylinders	In-line 4		V6	
Swept volume (cm^3)	1994		2494	
Bore (mm)	84.5		84.0	
Stroke (mm)	89.0		75.0	
Compression ratio	10:1		9.6:1	
Fuel grade (RON)	95, leaded or unleaded		97, leaded	
Injection system	Single point	Multi-point	Sequential	Multi-point
Power (kW)	88	103	127	123
at Speed (rpm)	5600	6000	6000	6000
Max. torque (Nm)	162	178	217	222
at Speed (rpm)	3500	4500	5000	4000
Valve timing:				
Inlet open (°btdc)	14	18	12.5	10
Inlet close (°abdc)	50	46	35	25
Exhaust open (°bbdc)	50	52	40	30
Exhaust close (°atdc)	14	12	12.5	10

final compromise was for the inlet tracts to be almost horizontal, and to converge slightly. During the induction process, barrel swirl (rotation about an axis parallel to the crankshaft) is produced in the cylinder. The reduction in volume during compression firstly causes an increase in the swirl ratio through the conservation of the moment of momentum. Subsequently, the further reduction in volume causes the swirl to break up into turbulence. This then enables air/fuel ratios of 18:1 ($\phi = 0.8$) to be burnt, thereby giving good fuel economy and low emissions.

The M16E engine uses a single-point fuel injection system (injection pressure 1 bar), and the M16I engine uses a multi-point fuel injection system (injection pressure 2.5 bar). The multi-point injection system uses one injector per cylinder, with the injector located in each inlet tract, upstream of the division into the two inlet ports. The air mass flow rate is deduced from the cooling effect on a heated wire. Other inputs to the computer-based control system include engine temperature, engine speed, throttle position and vehicle speed. The fuel supply is cut off on over-run in order to reduce the fuel consumption, in the manner described in section 1.3.

The ignition system is computer controlled, with the ignition timing map stored in memory as a function of the engine speed and load. The ignition timing is also retarded when knock occurs. Information from a knock sensor can be combined with crankshaft position information, to identify the cylinder in which knock is occurring. The ignition timing can then be retarded for that cylinder alone, until knock disappears. The benefits of this type of system were discussed in section 2.4.2.

As well as benefiting the engine operation, computing techniques have also helped during engine design and manufacture. Computer-aided design techniques were used to design the gas flow passages in the cylinder head. Computer-aided manufacture ensures that the important accuracy of these surfaces is maintained in manufacture, and flexible manufacturing automation enables different cylinder heads (including those of a high-speed direct injection diesel engine) to be machined using the same equipment.

The Rover 820i is fitted with either a five-speed manual gearbox, or a four-speed automatic gearbox. Since a different final drive ratio is used with each gearbox, table 7.3 lists the overall reduction ratios for each gearbox, as well as the gearbox reduction ratios.

With each gearbox, the step-up ratio between the gears reduces towards the top gear, for the reasons explained in section 4.2.4. The span of the gear ratios (the difference between the ratios in the highest and lowest gears) is 3.83 for the manual gearbox and 3.264 for the automatic gearbox. These spans are comparable with those of the example in table 4.5. The automatic gearbox can have a lower span, since the torque converter can magnify the torque output from the engine, thereby extending the span of the gearbox. The automatic gearbox used on the Rover 820 is the 4HP14 gearbox

Table 7.3 Rover 820 gearbox specifications

Gear no.	Manual gearbox			Automatic gearbox		
	Gearbox ratio	Overall ratio	km/h per 1000 rpm	Gearbox ratio	Overall ratio	km/h per 1000 rpm
1	3.250	12.795	9.06	2.412	10.625	10.91
2	1.894	7.457	15.55	1.369	6.030	19.23
3	1.307	5.146	22.53	1.000	4.405	26.32
4	1.033	4.067	28.51	0.739	3.255	35.62
5	0.848	3.339	34.72			
Reverse	3.000	11.811	9.82	2.828	12.457	9.31
Final drive	3.937			4.405		

manufactured by ZF, and the maximum torque multiplication of the torque converter is at least 1.9.

The 4HP14 gearbox has an epicyclic gear train, with hydraulic control of the clutches and brake bands to effect the different gear ratios. As can be seen from figure 7.4, the gearbox is very similar to the four-speed automatic gearbox illustrated in figure 4.9; the most substantial difference is the inclusion of the differential in the 4HP14, since it is for a front wheel drive application. In first, second and reverse gears, the drive is transmitted through the torque converter. The 4HP/4 gearbox also has a shaft driven directly by the engine, and in third gear only about 40 per cent of the power is transmitted through the torque converter. In fourth gear, the torque converter is bypassed, thereby avoiding the power dissipation associated with torque converter slip. The performance of the Rover 820E with the M16I engine and 4HP14 gearbox will be discussed after the treatment of the aerodynamic design and rolling resistance in the next section.

7.2.3 The Rover 800 body design and manufacture

The Rover 800 and Honda Legend have been designed around a common

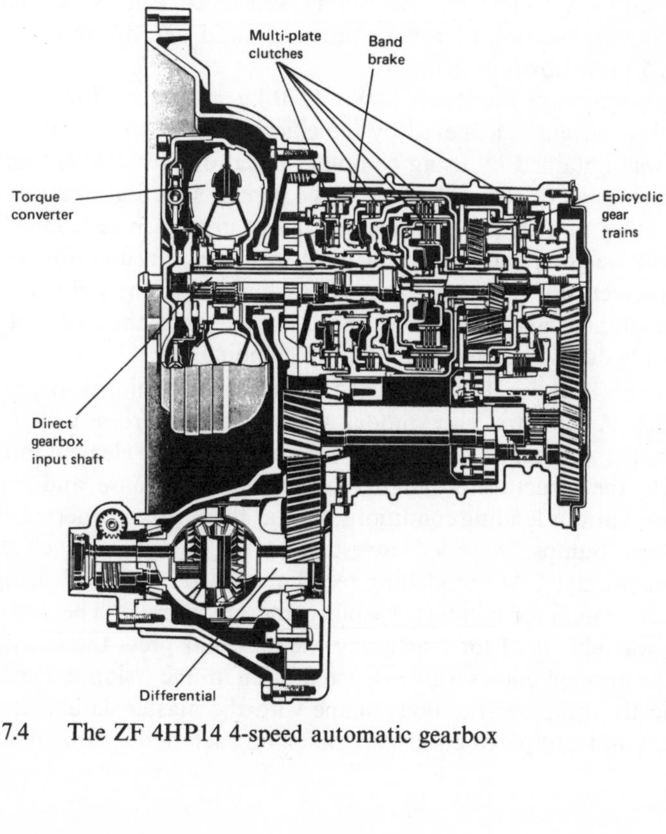

Figure 7.4 The ZF 4HP14 4-speed automatic gearbox

floor plan, engine bay, windscreen frame and most of the monosides. The full-size clay models of both vehicles were produced at the Austin Rover design studio, to ensure compatibility of the inner body panels and their welding.

The Rover 800 has a drag coefficient (C_d) of 0.32, and for the size of the car, a low frontal area of 2.03 m^2. The drag coefficient could have been reduced to 0.30 by adopting significant curvature of the bodysides. However, this would have adversely affected the position of the centre of pressure, and this would have produced poor stability in sidewinds. The lift coefficient was also subject to strict attention during development; the lift coefficient of 0.21 comprises the front lift coefficient (C_{lf}) of 0.15 and the rear lift coefficient (C_{lr}) of 0.06.

The good aerodynamic performance is a consequence of adopting the techniques described in section 5.3.2, coupled with extensive wind tunnel testing. The front air dam, bumper and nose are a carefully integrated design, and the air flows through the radiator and under the car are tightly controlled. The underfloor protrusions are minimised, and the door mirrors have also been carefully evolved to minimise drag. The traditional roof guttering has been eliminated and flush glazing has been adopted; both of these measures reduce the drag. By assuming typical values for the vehicle weight and the rolling resistance coefficient, the rolling resistance can be estimated as 175 N. The combination of rolling resistance and aerodynamic resistance (equation 5.1) is shown in figure 7.5.

The kerb weight of the Rover 820 is 1270 kg, and this is low for the size of car. A low weight is achieved by the efficient use of structural materials, and this was obtained by using computer-aided design (CAD) techniques. Weight savings were also obtained by the careful selection of materials for trim and other non-structural elements. It was argued in section 6.3 that a 20 per cent weight reduction would lead to a fuel consumption saving of about 8 per cent; of this, 4 per cent is attributable to the reduction in the rolling resistance, and the remainder is saved from the reduced power requirements during acceleration and hill climbing.

The Rover 800 has been designed using the techniques described in section 6.4. A full-size clay model was produced from which digital measurements were taken. These data were used for finite element modelling, to validate the structural strength, the dynamic response and for crash simulation. Various loading conditions, such as those from cornering, braking and vertical bumps, were all investigated to predict the vehicle ride. Three-dimensional CAD modelling techniques were also used to optimise the interior design, for comfort, visibility and ergonomics. The body shape database was also used for machining the dies that press the body panels. Finally, the finished bodies were inspected by an in-line vision test cell, which automatically compares the body shape with the master database. A total of 96 direct and computed checks are taken on each body, with an accuracy

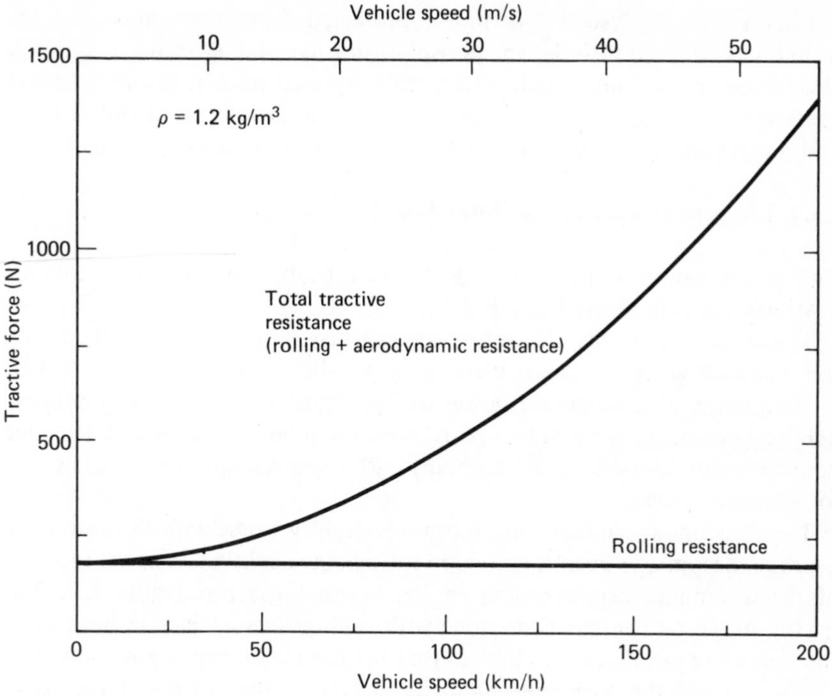

Figure 7.5 The rolling and aerodynamic resistance of the Rover 800

of 0.1 mm, by 62 laser cameras; the permitted tolerance in the body dimensions is 1 mm.

High-strength alloy mild steel is used for over 20 per cent of the body, and this provides a 13 per cent weight saving in these components. Bake-hardenable mild steel is used for the bonnet and boot; since its ductility is almost as good as that of mild steel, only low press tool loads are needed. The bake-hardenable mild steel then attains its high yield strength at the comparatively low temperatures of the paint ovens, and these panels then have a greater dent resistance. The bumpers are another example of a weight saving achieved by the combination of modern materials. A high-strength steel armature (with a yield strength increased by a factor of 6 through Nitrotech heat treatment) is covered by polyurethane foam with an outer skin of polybutadiene tetraphthalate (PBT). The PBT has a good impact resistance and an excellent surface for painting; the result is a bumper with half the weight of a conventional design. The flush glazing makes two contributions to vehicle design. Firstly, the aerodynamic performance is better and, secondly, glass that is secured with adhesives enhances the stiffness of the bodyshell.

Finally, the bodyshell has to be protected from corrosion, and the techniques adopted include zinc phosphatising, chromate rinsing, a cathodic electro-coat primer and an electrostatically applied primer, before the finish coats of paint are applied. Wax injection is used to protect the enclosed areas, and a urethane PVC mixture is used to protect the underside of the body.

7.2.4 The performance of the Rover 800

The performance of the Rover 820i, with both manual and automatic gearboxes, is summarised in table 7.4.

The acceleration with the automatic gearbox is slightly worse than with the manual gearbox, since there is a smaller number of gear ratios. Consequently, the maximum power of the engine is not so readily utilised, and furthermore in the first two gears, power will be dissipated by the torque converter slip. The urban fuel economy with the automatic gearbox is worse for similar reasons.

The constant-speed fuel consumption is slightly worse with the automatic gearbox, despite the overall reduction ratio being slightly less in the top gear of the automatic gearbox than in the manual gearbox (table 7.3). The maximum speed is almost the same with each gearbox, since in both cases the operating point is very close to that for the maximum engine power.

To estimate the 820i automatic performance, the tractive force curve (figure 7.5) has to be mapped on to the fuel-consumption map of the engine. The methodology for this mapping was developed in section 4.2.1. The ratio between the vehicle speed and engine speed is defined by the gearing ratios in table 7.3. The conversion of the tractive force (F) to the engine torque (T) depends on the transmission efficiency (η) as well as the gearing ratio (gr). For convenience, a value of 80 per cent will be assumed for the transmission

Table 7.4 Rover 820i performance with manual and automatic gearboxes

	820i manual	820i automatic
Acceleration (s):		
0–50 km/h	3.3	–
0–80 km/h	6.6	–
0–100 km/h	9.4	10.8
Maximum speed (km/h)	203	195
Fuel consumption (l/100 km):		
Urban cycle	10.5	12.8
90 km/h	6.6	6.9
120 km/h	8.2	8.5

efficiency. This low value has been chosen to include the power dissipated by the engine driven ancillaries, such as the power steering system and the alternator. The performance is analysed here for the 4th (top) gear, which will be seen later not to be an overdrive ratio in the strict sense.

$$\text{gearing ratio } (gr) = 35.62 \text{ km/h per 1000 rpm}$$

$$= \frac{35.62 \times 10^3}{60 \times 60} \bigg/ \frac{1000}{60}$$

$$= \frac{35.62}{60} = 0.594 \text{ m/rev.}$$

$$= 0.594/2\pi = 0.0945 \text{ m/rad}$$

By energy transfer

$$T = \frac{F \times gr}{\eta} = \frac{0.0945}{0.8} F$$

thus

$$T = 0.118F(\text{N}) \text{ N m}$$

The tractive force curve can now be superimposed as the road load curve on the fuel consumption map in figure 7.6. The intersection of the road load curve in top gear with the maximum torque envelope occurs at 5650 rpm, and this implies a maximum speed of 201 km/h, a value that is in close agreement with the speed of 195 km/h quoted in table 7.4.

On figure 7.6, the operating point for a speed of 90 km/h is marked as point A. This corresponds to a speed of 2527 rpm, a torque of 49.4 N m and, by interpolation, a specific fuel consumption of 335 g/kWh.

$$\text{Power, } \dot{W}_b = T \times \omega = 49.4 \times \frac{2527 \times 2\pi}{60} = 13.07 \text{ kW}$$

$$\text{Fuel flow rate } \dot{m}_f = \dot{W}_b \times \text{sfc} = 13.07 \times 0.335 = 4.38 \text{ kg/h}$$

This distance covered in 1 hour is 90 km, and thus the fuel consumption is

$$4.38 \times 100/90 = 4.87 \text{ kg/100 km}$$

If the density of petrol is 795 kg/m^3, then the fuel consumption is

$$5.87/0.795 = 6.12 \text{ l/100 km}$$

The operating point for 120 km/h (point B on figure 7.6) corresponds to a speed of 3669 rpm, a torque of 71.85 N m and, by interpolation, a specific fuel consumption of 298 g/kWh. Following the same procedure as above, the fuel consumption can be calculated as 7.92 l/100 km. These fuel consumptions are in quite close agreement with the values quoted in

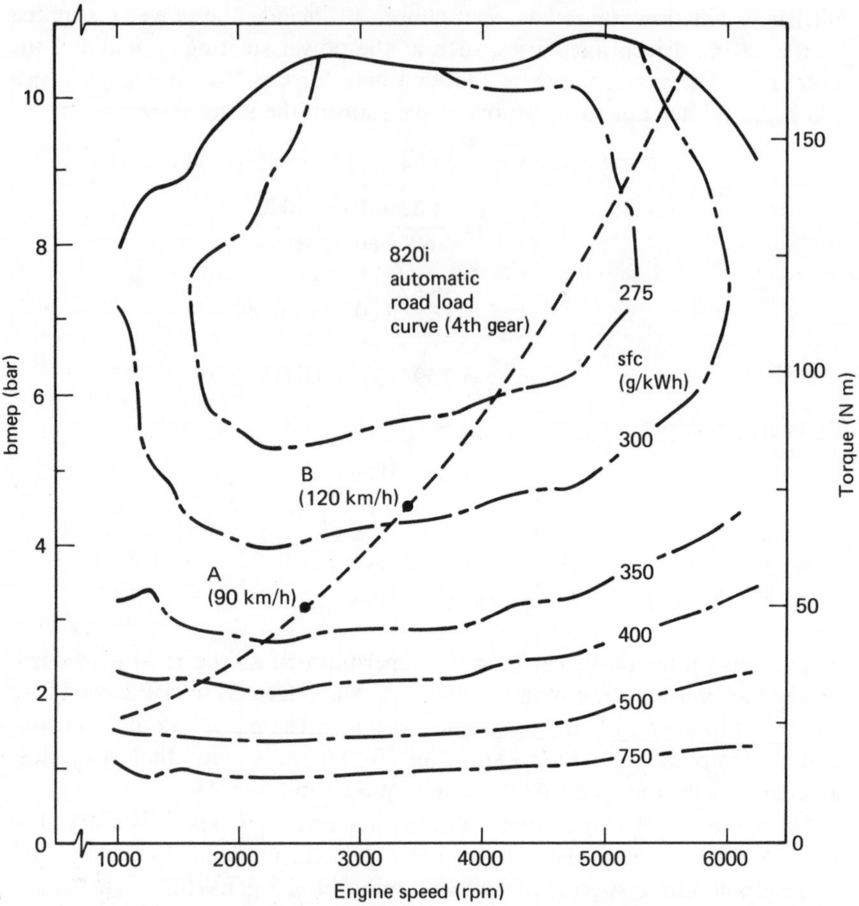

Figure 7.6 Specific fuel consumption map for the M16I engine, and the road load curve for the Rover 820i automatic in top (4th) gear

table 7.4.The calculated values are slightly low, and this suggests that the power dissipated by the transmission and the power consumed by the ancillaries have been underestimated at part load operation.

It should be noted that top gear in this vehicle is not strictly an overdrive ratio. The maximum speed is attained in top gear, but a higher speed could be obtained if the reduction ratio were slightly greater. Figure 7.6 shows that the specific fuel consumption is better than 300 g/kWh, for a major part of the engine map. By inspection, it can be seen that an overdrive ratio would not significantly increase the extent of the road load curve that fell within the 300 g/kWh contour. Consequently, the benefits of an overdrive ratio (with a slightly reduced efficiency) would be marginal.

7.3 The Ford Transit

7.3.1 Introduction

The Ford Transit van is a range of vehicles for the European market, with 37 different body styles and payload capacities in the range of 800–1900 kg. In addition to the various wheelbase lengths, roof heights, payloads, levels of trim and body types, each vehicle can have a range of powertrain components. Four different engines are used: 1.6 litre and 2.0 litre four cylinder spark ignition engines, a 2.9 litre V6 spark ignition engine, and a 2.5 litre direct injection diesel engine. Six different gearboxes are used: three four-speed gearboxes (one with an overdrive option), a five-speed gearbox, and a three-speed automatic gearbox. The final drive system includes four different tyre sizes and four different rear axles, three of which have different reduction ratios available.

This introduction should have been sufficient to show that compared with cars, vans have a much wider range of models. For simplicity, one particular vehicle will be considered in this case study: the Transit 100 short wheel base, low body van, with the 2.5 litre direct injection diesel engine and a five-speed manual gearbox. The designation '100' means a nominal payload of 1000 kg (in fact, the payload is 1100 kg). A van of this type is illustrated in figure 7.7 – although this particular illustration shows a four cylinder petrol engine.

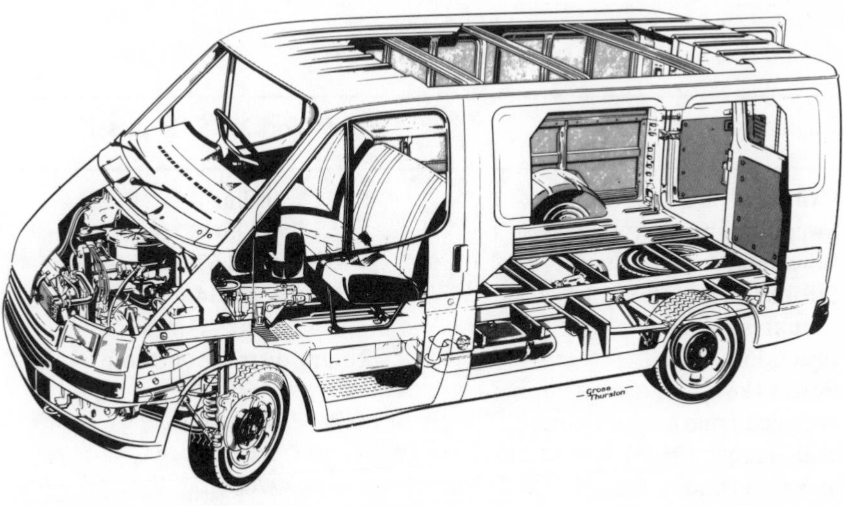

Figure 7.7 The Ford Transit short wheelbase van. [Reproduced by permission of Ford Motor Co. Ltd]

As with the Rover 800, a range of CAE techniques have been used for the design and manufacture of the Ford Transit. Perhaps the most remarkable aspect is the automated body production system. Robot welding lines are capable of producing 300 vehicles a day, with any sequence of different body styles. Aspects of the Transit development have been described by Garrett (1986) and Scott (1986b).

7.3.2 The Ford Transit powertrain

The 2.5 litre direct injection (DI) diesel engine that is being discussed in this case study is shown in figure 7.8. The differences between direct and indirect injection engines were discussed in section 3.3. To summarise, the efficiency of direct injection engines is some 10–15 per cent better than indirect injection engines, primarily because of the high heat transfer associated with the pre-chamber of IDI engines. The development of the Ford DI diesel has been described by Bird (1985), and the main parameters are summarised in table 7.5.

The cylinder block and head are both made from cast iron, with the cylinder axis inclined at 22.5° to the vertical. The cylinder spacing is the same as the now superseded indirect injection engine, so that the transfer machines can still produce replacement cylinder blocks for the old engine. The inlet and exhaust valves are in-line, but off-set from the engine centreline.The valves are operated by rocker arms driven by short push rods. The chilled iron camshaft is driven by a toothed rubber belt, which also drives the fuel injection pump. The inlet valve period (opening duration) is 232°, and the exhaust

Table 7.5 The Ford 2.5 DI diesel engine

Engine type	Ford 'York', 2.5 DI diesel
Cylinders	In-line 4
Swept volume (cm^3)	2496
Bore (mm)	93.67
Stroke (mm)	90.54
Compression ratio	19.02:1
Injection system	Rotary distributor pump
Power (kW)	50
at Speed (rpm)	4000
Max. torque (N m)	143
at Speed (rpm)	2700
Weight (kg)	219

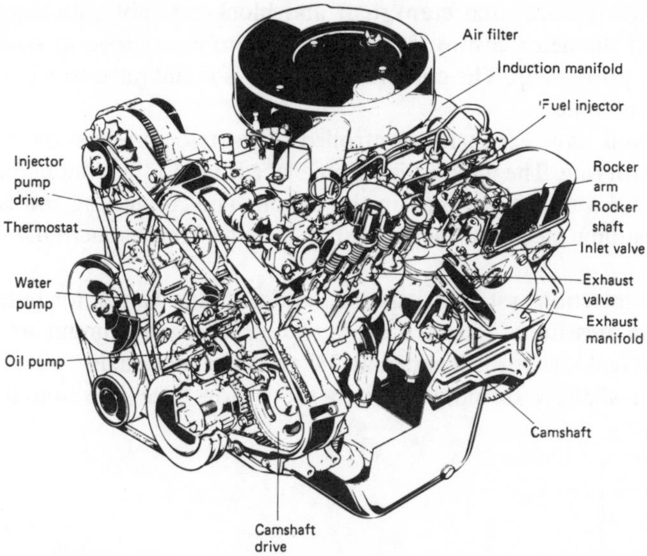

Figure 7.8 The Ford 2.5 litre DI diesel. [Reproduced by permission of Ford Motor Co. Ltd]

valve period is 244°; there is 26° of valve overlap. The valve guides are inserted into the cylinder head, and the flow of oil along the valve stems into the combustion chamber is controlled by polyacrylic seals. The exhaust valve stem is given a chrome finish, and the inlet valve stem is phosphated to give a long service life.

As explained in section 3.3, the satisfactory performance of high-speed direct injection engines depends on the careful matching of the air flow and fuel injection. In the first instance this requires extensive development work, and then during manufacture considerable care is needed to keep the engines within their tolerances. In the case of the Ford 2.5 DI engine, the induction system was designed with the use of CAD techniques, to define the inlet manifold and inlet port shapes. Engines with different swirl ratios were tested, to obtain the best compromise between swirl and volumetric efficiency. The inlet valve diameter and seat angle were also varied, before settling for an inlet valve seat angle of 30°, and an exhaust valve seat angle of 45°. The valve seats are cut directly in the head, and then induction hardened after machining.

During production, each assembled inlet manifold and cylinder head is checked for the correct swirl characteristics. Care is also needed in maintaining the combustion chamber geometry, especially the clearance between the piston crown and the cylinder head. Four grades of connecting rod length, and five grades of piston height (gudgeon pin to piston crown) permit selective assembly to the measured crankshaft and block assembly. Four grades of piston skirt diameter ensure the minimum bore clearances, to control the noise from piston slap. The compression ratio is maintained to within 0.7 of the nominal value.

The pistons have expansion controlled skirts, two compression rings and an oil control ring. The top ring is chrome faced, and for maximum durability it operates in a cast iron insert in the piston. The connecting rod is forged from a vanadium steel, and the crankshaft is made from a spheroidal graphite cast iron.

The fuel injection system is based around a rotary distributor pump, with a maximum pressure of 750 bar – about twice the usual maximum for a rotary pump. The fuel injector is inclined at 23° to the cylinder axis, and injects the fuel into a shallow toroidal bowl in the piston. The injection timing is

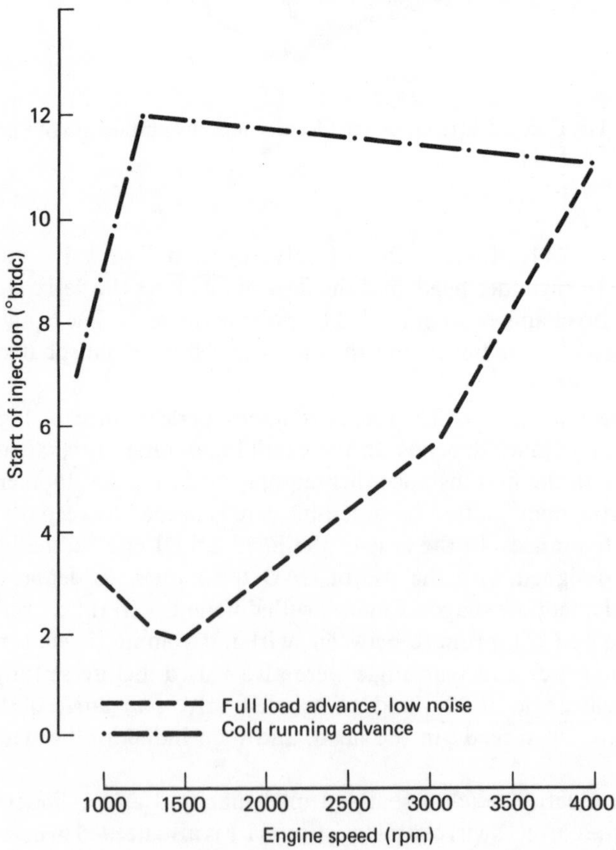

Figure 7.9 Dependence of the start of injection on engine speed and operating condition. [Reprinted by permission of the Council of the Institution of Mechanical Engineers]

controlled by a two-stage system, which increases the rate of advance above 3600 rpm. This system also allows the injection timing to be retarded at part load in the mid-speed range, in order to reduce the subjective noise levels. Under these conditions a careful balance is needed between emissions, noise and fuel economy (section 3.5). At coolant temperatures below 30°C, additional injection timing advance and a fast idle speed are used. These measures are to ensure good driveability, and to minimise the emissions of white smoke that are attributable to partially burnt fuel. The start of injection timing for two different operating conditions is shown in figure 7.9.

The maximum fuel delivery is also controlled as a function of speed, by a two-stage device. The comparatively high speed for the maximum torque is a consequence of limiting the fuelling at lower speeds, to avoid excessive emissions of smoke. This is a result of the swirl ratio needing to be higher at low speeds for the reasons presented in section 3.5.2. Evidently this fuel

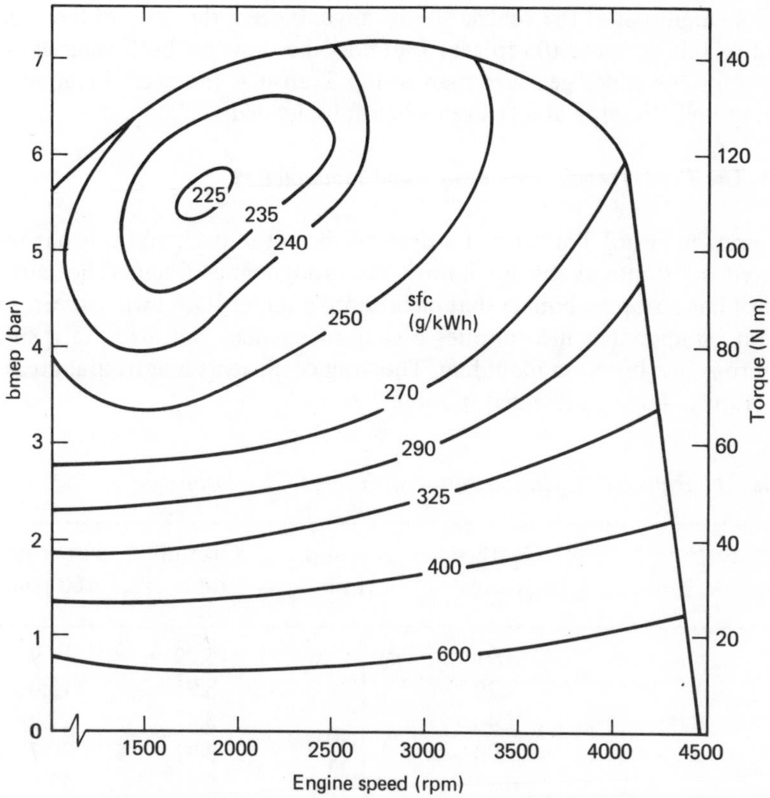

Figure 7.10 Specific fuel consumption contours (g/kWh) for the Ford 2.5 litre DI diesel engine. [Reprinted by permission of the Council of the Institution of Mechanical Engineers]

injection system has some very involved mechanical and hydraulic control systems, and the benefits in the future of adopting electronic control systems should be self-evident. The fuel economy of the Ford 2.5 DI diesel is shown in figure 7.10, and this performance is comparable with the Fiat engine shown in figure 3.4.

The five-speed gearbox that is being considered in this case study is designated the M27. The only obvious difference between this and the gearbox shown in figure 4.6 is that the M27 gearbox has the gear selector shaft mounted on the front cover of the gearbox. For the Transit 100 with the DI diesel engine and five-speed gearbox, a 3.91:1 reduction ratio is used in the type 34 rear axle, with 185R14 tyres. The results of this drivetrain combination are presented in table 7.6.

As with the gearboxes that have already been discussed, the M27 step-up ratio between the gears reduces towards top gear. The gearbox span is 4.77, and this is significantly greater than for the Rover 800 (3.83). While the speed range of 2.5 DI diesel is significantly smaller than the Rover M16 spark ignition engine, so is the vehicle speed range. Indeed, the ratio of the engine speed to vehicle speed in top gear is almost the same for both vehicles. The reason for the wider gearbox span in the Transit is the need for good hill starting and climbing ability even when fully loaded.

7.3.3 The Ford Transit body design and manufacture

The previous Ford Transit had a drag coefficient of 0.43, and this has been reduced substantially by an improved aerodynamic design. The current Transit has a sloping bonnet that blends into a direct glazed windscreen, and a front bumper that incorporates a shallow spoiler – all made in a single polypropylene injection moulding. The drag coefficients and frontal areas of the Transit 100 are presented in table 7.7

Table 7.6 Transit 100 gearing ratios with the M27 gearbox

Gear no.	Gearbox ratio	Step-up ratio	Overall ratio	km/h per 100 rpm
1	3.91		15.29	7.9
2	2.29	1.71	8.95	13.4
3	1.40	1.64	5.47	21.9
4	1.00	1.40	3.91	30.7
5	0.82	1.22	3.21	37.5
Reverse	3.66		14.31	8.4
Rear axle ratio	3.91		–	–

Table 7.7 Aerodynamic performance of the Transit 100

Vehicle	Drag coefficient C_d	Frontal area A (m^2)	C_dA (m^2)
Transit 100	0.37	3.34	1.237
Transit 100 (high roof)	0.35	3.73	1.306

The lower drag coefficient of the high-roof van is most likely to be due to the more curved transition from the windscreen to the roof. The aerodynamic resistance can be calculated from equation 5.1, and the rolling resistance can be calculated by assuming a rolling resistance coefficient (defined in section 6.2.2) of 0.012. The tractive resistance of the Transit 100 van is shown in figure 7.11 for both the unladen (1500 kg) and laden (2600 kg) cases. The vehicle performance with the 2.5 DI diesel engine will be assessed in the next section (7.3.4)

The bodyshell weight is 415 kg for the short wheelbase and 483 kg for the long wheelbase van; the design was optimised by finite element modelling techniques. Other examples of weight savings come from the innovative use of materials. The blow-moulded high-density polyethylene fuel tank provides a 5 kg weight saving over a steel tank. The use of an aluminium radiator with plastic end tanks gives a weight saving of over 25 per cent in comparison with conventional copper radiators.

Weight savings benefit the operator in terms of either reduced fuel consumption or by allowing a heavier payload. The best payload is achieved by the 2.0 litre spark ignition engined Transit, with a payload of 2010 kg, including the driver.

As stated in the introduction, the body construction plants each have a capacity of 300 vehicles a day, with the robot lines capable of assembling any sequence of different bodyshells. The body construction starts with the underbody, which is one of four different designs. Six gantry-mounted robots secure the longitudinal and cross members with 334 welds to the underbody, to construct the ladderframe. A further 32 robots make another 1460 welds to incorporate additional components, and to complete the underbody.

Similar robot lines produce the body sides and roof, ready for assembly to the underbody in the body framing line, where up to 1400 welds are made by a further 37 robots. A vision system uses five-cameras to check that the correct specification body is being built. Prior to painting, the body shells are cleaned, degreased and etched by a zinc phosphate process that is stabilised by a chromate rinse. The bodies are immersed in a tank for the cathodic electrocoating of the primer: this technique is chosen to ensure a

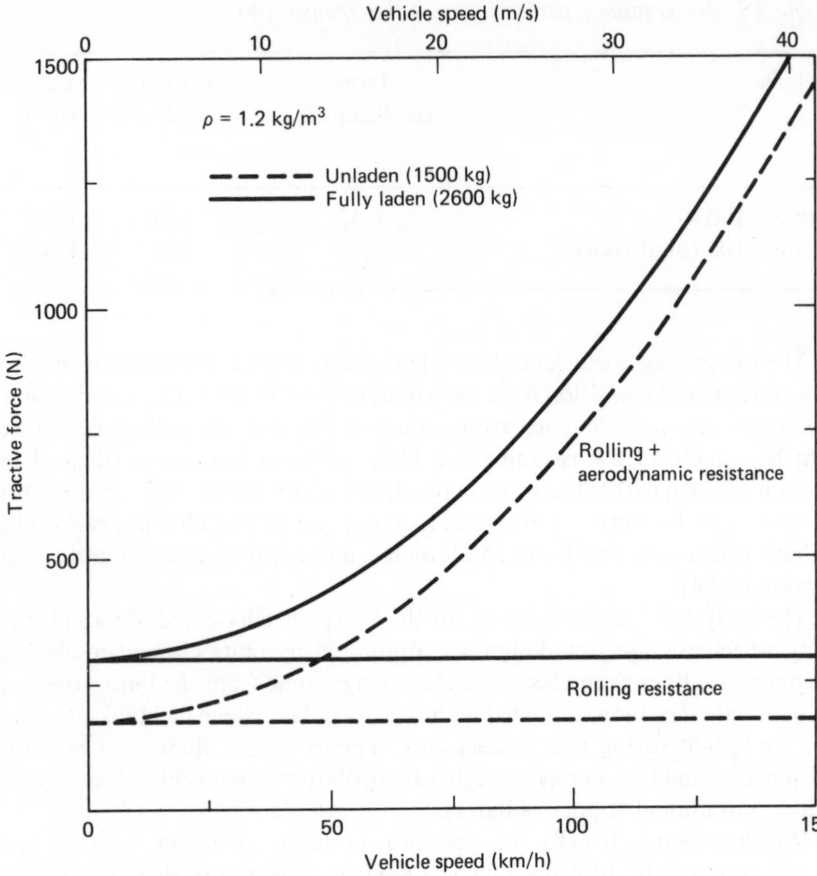

Figure 7.11 The rolling and aerodynamic resistance of the Ford Transit 100

uniform coating thickness in all areas. A polyester primer is applied electrostatically and then oven baked; two coats of enamel are applied 'wet on wet' and then oven dried. Additional measures to prevent corrosion are the chip-resistant PVC-based underseal applied to the wheel arches, and the high-pressure injection of wax, to coat the inner surfaces of all the cavities.

7.3.4 Performance of the Ford Transit 100

The vehicle that is the subject of this case study is the Transit 100, with a 2.5 litre DI diesel engine, a five-speed manual gearbox, and a short wheel base, standard roof height, body. The performance data produced by Ford for this vehicle are summarised in table 7.8.

The engine performance is defined by figure 7.10, and the drivetrain ratios

Table 7.8 Ford Transit 100 performance data (2.5 DI engine)

	Half payload	Full payload
Total weight (kg)	2050	2600
Acceleration (s):		
0–50 km/h	8.6	10.5
0–80 km/h	20.4	25.9
0–100 km/h	35.0	45.8
Maximum speed (km/h)	124	122
Fuel consumption (l/100 km):		
Urban cycle (kerbweight + 100 kg)	8.3	–
90 km/h	6.3	–
1st gear gradeability (%)	–	22.9

are presented in table 7.6. The traction requirements are defined by figure 7.11, and in order for the vehicle performance to be estimated, these data have to be mapped on to figure 7.10. As in the previous case study (section 7.2.4), the power absorbed by the ancillaries, such as the alternator and vacuum pump for the brakes, will be treated as part of the transmission losses, and a value of 80 per cent will be used for the overall transmission efficiency.

Consider first the maximum vehicle speed; this will be attained in gear 4. The overall gearing ratio of 30.7 km/h per 1000 rpm allows direct scaling between the engine speed and vehicle speed. The conversion from tractive force (F) to the engine torque (T) is slightly more involved, since the transmission efficiency (η) needs to be included as well as the gearing ratio (gr); the approach is the same as that developed in section 4.2.1.

$$gr = 30.7 \text{ km/h per 1000 rpm}$$

$$= \frac{30.7 \times 10^3}{60 \times 60} \bigg/ \frac{1000}{60} \text{ m/rev.}$$

$$= 30.7/60 = 0.512 \text{ m/rev.}$$

$$= 0.512/2\pi = 0.0814 \text{ m/rad}$$

By energy transfer:

$$T = \frac{F \times gr}{\eta} = \frac{0.0814}{0.8} F \text{ N m}$$

thus

$$T = 0.1018F(\text{N}) \text{ N m}$$

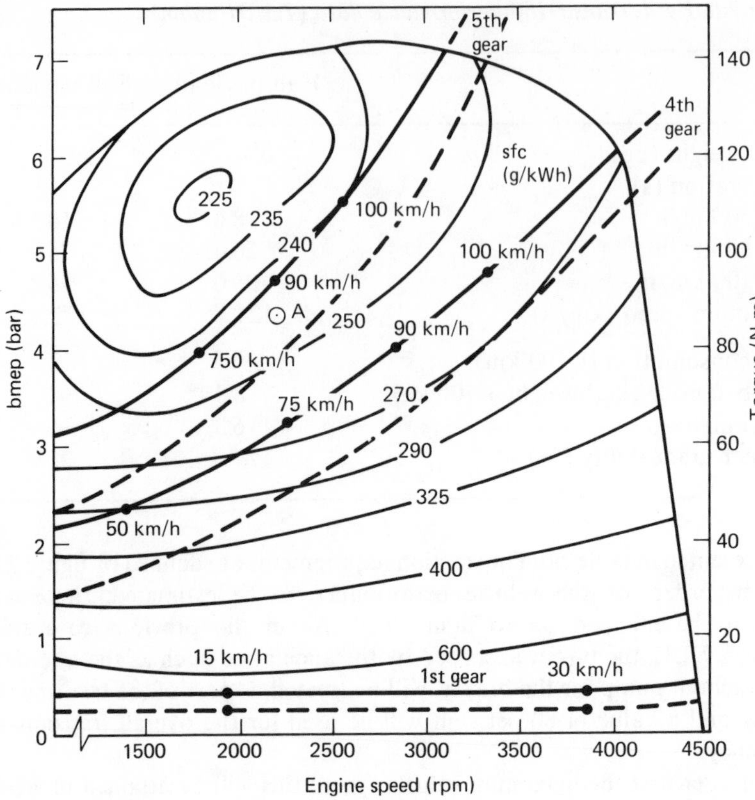

Figure 7.12 Road load curves for a Transit 100 van in 4th and 5th gears, at full load (——————), and no load (– – – –)

The tractive force data from figure 7.11 can now be mapped on to figure 7.10, and the process can be repeated for the 5th gear, to produce figure 7.12.

The intersection of the road load curve in 4th gear with the engine torque envelope gives the engine speed at the maximum vehicle speed. For full load this corresponds to 122 km/h, and for half load this corresponds to 123 km/h; these values agree well with those presented in table 7.8. The corresponding maximum speeds in 5th gear are 116 km/h and 124 km/h.

On figure 7.12, point A represents the operating point for 90 km/h with half payload. This corresponds to an engine speed of 2400 rpm, a torque of 86.3 N m and, by interpolation, a specific fuel consumption of 245 g/kWh.

$$\text{Power, } \dot{W}_b = T \times \omega = 86.3 \times \frac{2400}{60} \times 2\pi = 21.7 \text{ kW}$$

$$\text{Fuel flow rate, } \dot{m}_f = \dot{W}_b \times \text{sfc} = 21.7 \times 0.245 = 5.31 \text{ kg/h}$$

The distance covered in one hour is 90 km, thus the fuel consumption is (5.31/0.9) = 5.9 kg/100 km.

If the density of diesel fuel is taken as 920 kg/m^3, then the fuel consumption is (5.9/0.92), or 6.4 l/100 km; this is in good agreement with the value of 6.3 l/100 km in table 7.8. The hill-climbing ability, or gradeability, is found by considering the maximum engine torque that can be applied through first gear.

Maximum engine torque (T) is 143 N m

Gearing ratio in 1st gear (gr) is 7.9 km/h per 1000 rpm
$\equiv 0.0208$ m/rad

Efficiency (η) is 80 per cent

The maximum tractive effort available, F, is thus

$$F = \frac{T \times \eta}{gr} = \frac{143 \times 0.8}{0.0208} = 5.50 \text{ kN}$$

The corresponding engine speed is 2700 rpm, at which the vehicle speed would be 21.3 km/h, and the tractive resistance on level ground would be 346 N.

$$\text{gradeability} = \frac{F - 346}{Mg} \times 100 \text{ per cent}$$

$$= \frac{5500 - 346}{2600 \times 9.81} = 20.2 \text{ per cent}$$

Again this is in acceptable agreement with the value of 22.9 per cent quoted in table 7.8.

7.4 Conclusions to motor vehicle fuel economy

Good motor vehicle economy is obtained by treating the vehicle as a system, and ensuring that each element in the system is designed to meet the system specification as efficiently as possible. The performance of particular elements in the system can be compared with those of equivalent elements in other vehicles. Any element that compares unfavourably (thereby constituting a weak link in the chain) should be appraised to see if any modifications would lead to a useful gain in fuel economy.

The first requirement is to produce the power as efficiently as possible from an available fuel, while remaining within acceptable emissions limits. With any engine, it is very rarely the maximum efficiency that is important. Engines invariably operate at part load, and normally the operating point is not the most efficient available for a given power output. Even a continuously variable transmission cannot always optimise the operating point, if it has a finite gear ratio span. The part load efficiency of any engine is very dependent on the power requirements of ancillary devices and the mechanical efficiency of the engine; these topics were discussed in section 2.3.

The efficiency of spark ignition engines is influenced most strongly by the compression ratio and the air/fuel ratio (section 2.2). A high compression ratio and a weak air/fuel mixture will give the best fuel economy, but will require a carefully designed combustion chamber and a suitable (high octane rating) fuel. In the past, the octane rating of fuels has been increased by organic lead additives, but this is now substantially limited by legislation. High octane fuels require more energy for their manufacture, and when the engine and refinery are treated as a system, the optimum octane rating is about 95 RON for Europe [Dorgham (1982)]. The effects of the compression ratio and air/fuel ratio on emissions have been discussed in section 2.4.3, and the European legislation is summarised in table 2.2. It seems likely that small engines will meet the emissions requirements by control of the air/fuel ratio, while larger engines will also require exhaust gas catalysts. Whether the air/fuel mixture is prepared by a carburettor or by an electronic fuel injection system, adequate control is needed. It is also important that the inter-cylinder variation in air/fuel ratio is minimised – only one air/fuel ratio can be optimal at any one time.

The mixture also has to be ignited at the correct time, and as a minimum the timing needs to vary with the engine load and speed. The ignition system also needs to provide sufficient energy for reliable and consistent ignition; this will help to minimise emissions, the fuel consumption and cyclic dispersion – a phenomenon that is most prevalent in lean burn engines operating at part throttle.

In order to optimise and control the air flow over the wide speed range, which is characteristic of spark ignition engines, electronically controlled variable induction systems and variable valve timing systems are likely to be used. These have been described in section 2.4.1.

Any comparison between spark ignition and diesel engines, needs to consider the differences in fuel energies and production costs that are listed in table 3.1.The reasons why diesel engines are more efficient than spark ignition engines are presented on page 57. Furthermore, the fall in efficiency of diesel engines at part load is less severe than for spark ignition engines, as explained on page 58.

The speed range limitation of direct injection (DI) diesel engines has meant that the less efficient indirect injection (IDI) engines have been used in light vehicle applications. These two types of engine are compared in section 3.3. The IDI engine is less efficient, because of the increased heat transfer associated with a divided combustion chamber and the high compression ratio. Higher compression ratios are usually expected to improve efficiency, but the compression ratios of IDI engines are so high (usually over 20:1) that the increased mechanical losses (which accompany the higher pressure loadings) more than offset any thermodynamic gains. However, the speed range of DI engines is increasing, and the performance of a high-speed DI engine was compared with an IDI engine in figure 3.4. The fuel economy of a DI engine is better than an IDI by more than 10 per cent at almost all loads and speeds.

A major disadvantage of diesel engines is their poor specific output, but this can be improved by turbocharging. A turbocharged DI diesel can have an output of 28 kW/litre, but this is still low in comparison with the fairly readily achievable 40 kW/litre from naturally aspirated spark ignition engines. The use of a diesel engine can introduce a weight penalty of about 6 per cent in a typical automobile, with an associated fuel consumption penalty of 3–5 per cent. An added advantage of turbocharging diesel engines is the improved efficiency, which arises from the output increasing faster than the associated mechanical losses. Another means of improving the performance of diesel engines (which was also discussed in section 3.4) is the adoption of a low heat loss design. The reduced heat transfer increases the expansion work, and leads to higher exhaust gas temperatures. Furthermore, the reduced cooling requirements permit a smaller cooling system. The lower power requirement for the cooling systems gives an improvement in economy, which is most significant at part load.

The emissions from diesel engines (section 3.5.1) are inherently lower than those of spark ignition engines. However, the timing and rate of injection have to be optimised, in terms of their effect on noise and gaseous emissions, fuel economy and output. The engine emissions are also influenced by the engine load, speed, temperature, boost pressure, air inlet temperature, exhaust gas recirculation and fuel quality. Consequently, electronic fuel injection control systems are being developed to ensure optimised performance (section 3.5.2).

For good vehicle fuel economy, not only does the transmission system have to be efficient, but it also has to match the vehicle speed to the engine speed, so as to give an efficient engine operating condition (in general, a high load and a low speed). The other basic requirements are firstly, that the transmission should enable the vehicle to start on a steep gradient when fully laden and, secondly, that the potential maximum speed is achieved. The principles in powertrain matching were developed in section 4.2.1.

In general, the maximum vehicle speed occurs in direct drive through the gearbox, and this defines the rear axle reduction ratio for a given tyre rolling radius. The hill-starting requirement then defines the maximum reduction ratio required in the gearbox. The tractive effort or force curves (figure 5.12) can be mapped on to the engine map (figure 4.2) or the engine characteristics can be mapped on to the tractive force curves (figure 4.4). In either case, due account has to be taken of the efficiencies of the drive train elements. Unfortunately, with this simple approach, the road load curve does not coincide with the regions of high engine efficiency at typical vehicle speeds. This leads to the use of overdrive ratios, the effects of which were shown on figure 4.2 and discussed in section 4.2.2. The overdrive ratio (typically about 0.8:1) reduces the engine speed and increases the engine load, thereby achieving a more efficient operating point. In the example recorded in table 4.2, the improvements in efficiency through using an overdrive ratio were as high as 24 per cent. However, in normal driving, which is not at constant speed, the gains will be lower.

Continuously variable transmissions (sections 4.2.3 and 4.5) would appear to offer scope for improved fuel consumption. However, their efficiency is about 90 per cent of that of conventional transmissions, and their span is often limited (the ratio of the minimum to maximum reduction ratios available). When a continously variable transmission has a limited span, it means that the optimum operating point for a given power requirement cannot always be obtained (figure 4.3). Furthermore, the reduced efficiency leads to a reduction in the maximum vehicle speed, and a poorer fuel consumption at high speeds (table 4.4).

Automatic transmissions (section 4.4) have traditionally been associated with a poor vehicle performance and a high fuel consumption. However, the development of four-speed automatic gearboxes with overdrive ratios, and torque converters with lock-up, has led to automatic transmissions with performances comparable with those of manual transmissions. Torque converter slip would otherwise lead to a direct increase in fuel consumption: 5 per cent slip would imply a 5 per cent loss of energy. Another significant source of power dissipation in automatic gearboxes is the hydraulic pump. By using a variable capacity pump, it has been reported that the power consumption is reduced by 1 kW.

Powertrain optimisation (section 4.6) is most easily achieved through the use of computer modelling. Different engine and transmission combinations can be analysed, to yield the desired trade-off between performance and economy (figures 4.17 and 4.18). Additional fuel savings can be achieved by regenerative braking; this is particularly attractive for vehicles with frequent stops and starts. Commercial systems have been produced for buses – these use kinetic energy stored in a flywheel or hydraulic energy stored in an accumulator – and fuel savings of up to 30 per cent have been claimed.

Aerodynamic resistance has its strongest influence on the vehicle fuel consumption at high speeds, and on the maximum vehicle speed. For a typical car, reducing the drag coefficient from 0.45 to 0.33 would give improvements in the constant-speed fuel consumption, ranging from 6.7 per cent at 40 km/h to 22.5 per cent at 160 km/h (table 5.2). The power for driving the vehicle with a drag coefficient of 0.33 at 160 km/h would only be sufficient to drive the vehicle at 145 km/h if the drag coefficient was 0.45 (figure 5.12). For trucks, the aerodynamic resistance is still significant (figure 5.18), especially when unladen. Aerodynamics is particularly important for coaches, since their speed is greater and their payload is smaller than for trucks. A good drag coefficient for a car would be about 0.3, while that of a van might be about 0.35, and the drag coefficient of a coach could be reduced to about 0.4. The control of drag in trucks is more difficult, especially in those with separate tractors and trailers. By rounding the front of the tractor and controlling the gap between the tractor and trailer, the drag coefficient can be reduced from greater than 0.7 to within the range of 0.4–0.5.

Vehicle aerodynamics is not solely concerned with drag reduction, since control of the centre of pressure and the lift forces is also essential. It should also be remembered that there will usually be a cross-wind present, which produces a net flow with a finite yaw angle. Thus the drag coefficient should not rise rapidly with non-zero yaw angles. Numerical techniques for predicting drag are improving, but it is still necessary to do extensive wind tunnel testing of models and full-size vehicles, in order to optimise the aerodynamic performance.

Vehicle design affects the rolling resistance, and thereby fuel economy, in two ways. Firstly, the design of the brakes, the wheel bearings and, most significantly, the tyres has a direct link with the vehicle rolling resistance. Secondly, any weight reduction leads to a direct reduction in the rolling resistance (the tyre rolling resistance coefficients are nearly constant at about 1–1.5 per cent). These measures produce direct savings in the steady-state fuel consumption; furthermore, weight reductions lead to less energy being needed for acceleration and hill climbing. When any weight savings are made in a component, weight compounding will lead to additional weight savings in other components. For example, if the mass of the engine is reduced by 10 kg, this will probably permit a further saving of about 5 kg in the remainder of the vehicle.

A 20 per cent reduction in the rolling resistance of a passenger car might lead to improvements in the steady-state fuel consumption ranging from 8 per cent at 40 km/h to 2 per cent at 160 km/h (table 6.1). However, the greatest potential for gains occurs at low speeds, but this is also where speed variations lead to the largest discrepancy from the steady-state fuel consumption predictions. In practice, a 20 per cent reduction in rolling resistance might lead to a 4 per cent improvement in fuel economy, a result that is almost independent of the driving pattern. In trucks, the rolling resistance is more significant (table 6.2); a 20 per cent reduction in rolling resistance would typically produce a 7 per cent reduction in the fuel consumption. The rolling resistance of tyres is affected by the road surface, tread pattern, the properties of the rubber, the inflation pressure and the tyre design. Radial tyres have a rolling resistance coefficient that is about 20–30 per cent lower than that of cross-ply tyres, and this accounts for their almost universal use on cars, and increasing use on trucks.

The speed variations of vehicles does mean that the savings in fuel consumption attributable to weight reduction are greater than those solely attributable to the reductions in the rolling resistance. A reduction of 20 per cent in the vehicle mass would lead to a fuel saving of somewhere between 8 and 15 per cent, of which about 4 per cent would be attributable to the 20 per cent reduction in the rolling resistance. In the case of trucks, any reduction in the vehicle mass is likely to be used as a means of increasing the payload of the vehicle. Weight savings are obtained by using more sophisticated design techniques (section 6.4), in combination with the use of a wide range of carefully chosen materials (section 6.5).

Motor vehicle fuel economy is also affected by vehicle usage patterns (section 1.2). Vehicles need to be maintained correctly to give a low fuel consumption, and in addition to this, the driving style also has a profound effect on the fuel consumption. In city driving, the road design is also significant. In West German cities, 35–40 per cent of the time can be spent idling, and a 20 per cent fuel saving has been atrributed to improved traffic flow. Additional techniques for improving urban fuel economy are a reduction of the idling speed, cutting the fuel supply during deceleration, and switching the engine off after a fixed duration wait. Another technique for improving the fuel economy in the real world is to reduce the fuel consumption penalty associated with engine warm-up. Many journeys start with a cold engine and are of a duration that is insufficient to obtain steady-state conditions (figures 1.3 and 1.4). As much as 10 per cent of the total fuel consumed may be a consequence of the reduced fuel economy during engine warm-up.

The preceding remarks should have illustrated how good fuel economy has been obtained, and how it is going to continue to improve. The most significant developments for cars will be the introduction of high-speed direct injection diesel engines and microprocessor-controlled transmissions. Further aerodynamic improvements are possible, but they will be influenced by public taste and manufacturing costs. In larger vehicles, the use of the so-called adiabatic (and turbocharged) diesel engines will lead to additional fuel savings. The multi-ratio gearboxes are also a particularly attractive application for microprocessor control. When there are as many as 16 gear ratios to choose from, an intelligent control system is the only way to ensure that the correct ratio will always be selected. To conclude then, a continued improvement in fuel economy can be expected in all classes of vehicle.

Appendix A: SI Units and Conversion Factors

SI (Système International) Units are widely used, and adopt prefixes in multiple powers of one-thousand to establish the size ranges. Using the watt (W) as an example of a base unit:

picowatt	(pW)	10^{-12} W
nanowatt	(nW)	10^{-9} W
microwatt	(μW)	10^{-6} W
milliwatt	(mW)	10^{-3} W
watt	(W)	1 W
kilowatt	(kW)	10^{3} W
megawatt	(MW)	10^{6} W
gigawatt	(GW)	10^{9} W
terawatt	(TW)	10^{12} W

It is unusual for any single unit to have such a size range, nor are the prefixes nano (10^{-9}) and giga (10^{9}) very commonly used.

An exception to the prefix rule is the base unit for mass – the kilogram. Quantities of 1000 kg and over commonly use the tonne (t) as the base unit (1 tonne (t) = 1000 kg).

Sometimes a size range using the preferred prefixes is inconvenient. A notable example is volume; here there is a difference of 10^{9} between mm^{3} and m^{3}. Consequently it is very convenient to make use of additional metric units:

$$1 \text{ cm} = 10^{-2} \text{ m}$$

thus

$$1 \text{ cm}^3 = 10^3 \text{ mm}^3 = 10^{-6} \text{ m}^3$$

$$1 \text{ litre (l)} = 1000 \text{ cm}^3 = 10^6 \text{ mm}^3 = 10^{-3} \text{ m}^3$$

Pressure in SI units is the unit of force per unit area (N/m^{2}), and this is sometimes denoted by the Pascal (Pa). A widely used unit is the bar

(1 bar $= 10^5$ N/m^2), since this is nearly equal to the standard atmosphere:

$$1 \text{ standard atmosphere (atm)} = 1.01325 \text{ bar}$$

A unit commonly used for low pressures is the torr:

$$1 \text{ torr} = \frac{1}{760} \text{atm.}$$

In an earlier metric system (cgs), 1 torr = 1 mm Hg.

The unit for thermodynamic temperature (T) is the kelvin with the symbol K (*not* °K). Through long established habit a truncated thermodynamic temperature is used, called the Celsius temperature (t). This is defined by

$$t = (T - 273.15)\,^\circ\text{C}$$

Note that (strictly) temperature differences should always be expressed in terms of kelvins.

Some additional metric (non-SI) units include:

Length	1 micron $= 10^{-6}$ m
	1 angstrom (Å) $= 10^{-10}$ m
Force	1 dyne (dyn) $= 10^{-5}$ N
Energy	1 erg $= 10^{-7}$ N m $= 10^{-7}$ J
	1 calorie (cal) $= 4.1868$ J
Dynamic viscosity	1 poise (P) $= 1$ g/cm s $= 0.1$ N s/m^2
Kinematic viscosity	1 stokes (St) $= 1$ cm^2/s $= 10^{-4}$ m^2/s

A very thorough and complete set of definitions for SI Units, with conversions to other unit systems, is given by Haywood (1972).

Conversion factors for non-SI Units

Exact definitions of some basic units:

Length	1 yard (yd) $= 0.9144$ m
Mass	1 pound (lb) $= 0.453\,592\,37$ kg
Force	1 pound force (lbf) $= \frac{9.806\,65}{0.3048}$ pdl
	(1 poundal (pdl) $= 1$ lb ft/s^2)

Most of the following conversions are approximations:

Length	1 inch (in) $= 25.4$ mm
	1 foot (ft) $= 0.3048$ m
	1 mile (mile) ≈ 1.61 km
Area	1 square inch (sq.in) $= 645.16$ mm^2
	1 square foot (sq.ft) ≈ 0.0929 m^2

Volume 1 cubic inch (cu.in.) ≈ 16.39 cm^3
1 gallon (gal) ≈ 4.546 l
1 US gallon ≈ 3.785 l

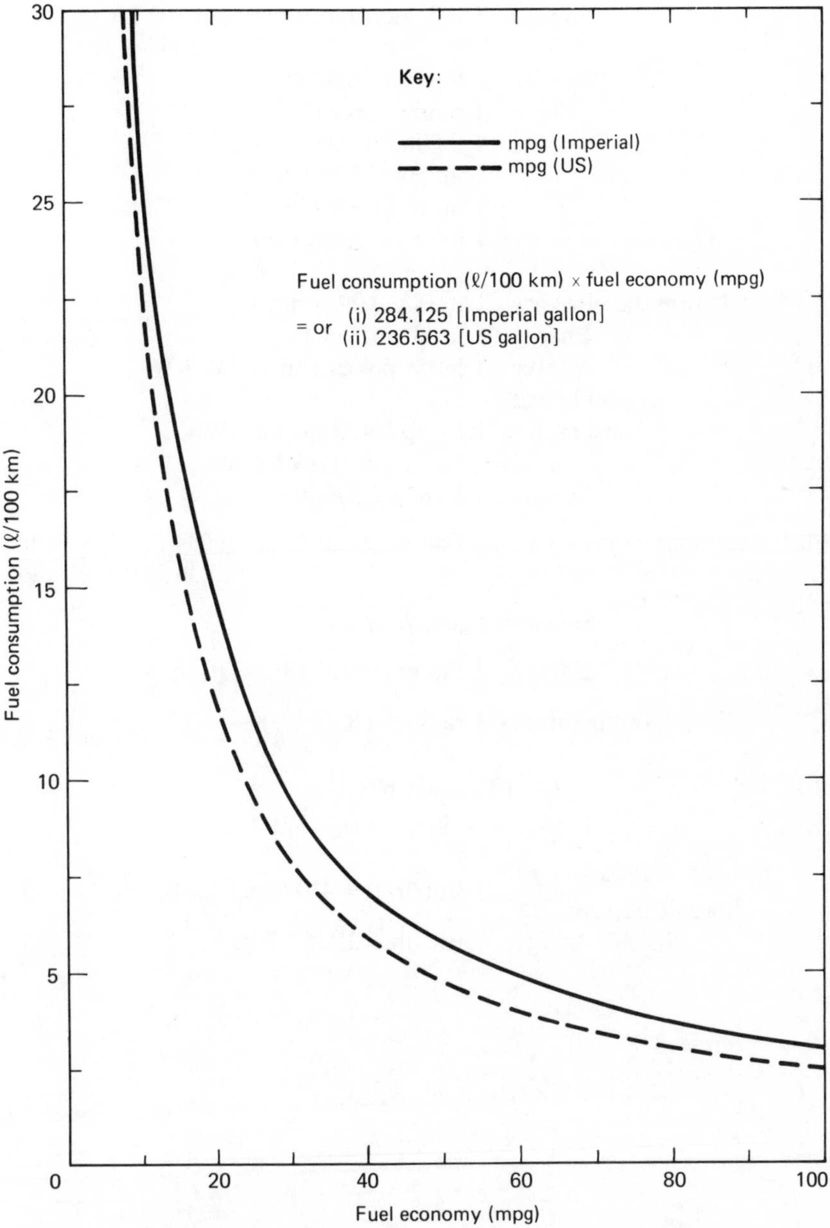

Figure A.1 Conversion graph for litres/100 km to miles per gallon

Mass	1 ounce (oz) ≈ 28.35 g
	1 pound (lb) ≈ 0.4536 kg
	1 ton (ton) ≈ 1016 kg
	1 US short ton ≈ 907 kg
Speed	1 mile per hour (mph) ≈ 1.609 km/h
	≈ 0.447 m/s
Density	1 $lb/ft^3 \approx 16.02\ kg/m^3$
Force	1 pound force (lbf) ≈ 4.45 N
Pressure	1 $lbf/in.^2 \approx 6.895\ kN/m^2$
	1 in. Hg ≈ 3.39 kN/m^2
	1 in. $H_2O \approx 0.249\ kN/m^2$
Dynamic viscosity	1 lb/ft s ≈ 1.488 kg/m s
	N s/m^2
Kinematic viscosity	1 $ft^2/s \approx 0.0929\ m^2/s$
Energy	1 ft lbf ≈ 1.356 J
Power	1 horse power (hp) = 745.7 W
Specific fuel consumption	1 lb/hp h ≈ 0.608 kg/kWh
	≈ 0.169 kg/MJ
Torque	1 ft lbf ≈ 1.356 N m

Fuel consumption (litres/1 km) × Fuel economy (mpg) ≈ 284.125 (UK gallon)
≈ 236.563 (US gallon)

(see also figure A.1)

Energy	1 therm (= 10^5 Btu) ≈ 105.5. MJ
Temperature	1 rankine (R) = $\frac{1}{1.8}$K

$$\left\{\begin{array}{l} t_F = (T_R - 459.67)^\circ F \\ \text{thus } t_F + 40 = 1.8\ (t_C + 40) \end{array}\right\}$$

Specific heat capacity, Specific entropy	1 Btu/lb R = 4.1868 kJ/kg K
Specific energy	1 Btu/lb = 2.326 kJ/kg

Appendix B: Rolling Radii of Radial Car Tyres

Radial tyres have a designation that indicates the width, construction, rim size, load index and speed rating. For example, 155R12 685 means a nominal width of 155 mm, of Radial construction to fit a 12 inch diameter wheel rim, with a load index of 68 and a speed rating of 5 (up to 180 km/h). A tyre with a 70 per cent aspect ratio would be designed 155/70 R12.

Tyre size	Effective rolling radius (mm)	Revolutions per kilometre	Revolutions per mile
145R10	237	671	1080
155/80R10	241	659	1061
125R12	248	641	1032
135R12	253	629	1012
145R12	263	606	975
145/70R12	250	637	1025
155R12	269	592	952
155/70R12	257	619	996
135R12	266	599	964
145R13	274	582	936
155R13	280	568	914
155/70R13	270	590	949
165R13	288	552	889
165/70R13	274	582	936
175R13	293	544	875
175/60R13	263	606	975
175/70R13	281	567	912
185R13	302	526	847
185/60R13	267	597	960
185/70R13	288	553	890
195/70R13	294	542	872
6.40R13	310	513	825

Tyre size	Effective rolling radius (mm)	Revolutions per kilometre	Revolutions per mile
7.25R13	315	505	813
135R14	278	572	920
145R14	287	556	894
155R14	295	541	870
165R14	301	529	851
165/70R14	287	556	894
175R14	307	518	834
175/65R14	281	567	912
175/70R14	291	546	879
185R14	314	508	817
185/60R14	279	570	917
185/65R14	288	552	889
185/70R14	300	530	853
195/60R14	285	558	898
195/70R14	305	521	839
205/70R14	311	511	823
215/70R14	321	495	797
125R15	293	544	875
135R15	293	544	875
145R15	301	529	851
155R15	307	518	834
165R15	315	505	813
175R15	323	493	793
175/70R15	306	521	838
180R15	328	485	781
185R15	333	478	770
185/65R15	302	528	849
185/70R15	317	503	809
195/60R15	297	535	861
195/70R15	321	495	797
205/60R15	305	522	840
205/70R15	327	487	783
215R15	352	452	727
215/70R15	334	476	766
225/70R15	341	467	752
235/70R15	343	464	747
205R16	357	446	717

Bibliography

The most prolific source of published material on motor vehicles is the Society of Automotive Engineers (SAE) of America. Some of the individual papers are selected for inclusion in the annual *SAE Transactions.* Other SAE publications include the *Progress in Technology* (PT) and *Specialist Publication* (SP), in which appropriate papers are grouped together. Examples are

SP-532 *Aspects of Internal Combustion Engine Design*
PT-27 *Engine-Oil Effects on Vehicle Fuel Economy*

The SAE also organise a wide range of meetings and conferences, and publish the monthly magazine *Automotive Engineering.*

In the United Kingdom, the Institution of Mechanical Engineers (I. Mech. E.) publish *Proceedings*, and hold seminars and conferences, many of which relate to vehicles and internal combustion engines. The Automobile Division also publishes the bi-monthly *Automotive Engineer.*

The other main organisers of European conferences include:

CIMAC	Conseil International des Machines à Combustion
FISITA	Fédération International des Sociétés d'Ingénieur et de Techniciens de l'Automobile
IAVD	International Association for Vehicle Design
ISATA	International Symposium on Automotive Technology and Automation

Finally, the JSAE (Japanese Society of Automotive Engineers) *Review* publishes translations of selected Japanese papers; any work that merits such translation often merits reading.

References

R. L. Abbot (1972), 'Overdrives', *Proc. I. Mech. E. 1969–1970*, Vol. 184, pt. 31, pp. 272–83, published as *Drive Line Engineering*, Book 1, I. Mech. E., London

A. Allard (1982), *Turbocharging and Supercharging*, Patrick Stephens, Cambridge

M. F. Ashby and D. R. H. Jones (1980), *Engineering Materials*, Pergamon, Oxford

J. Atkinson and O. Postle (1977), 'The effect of vehicle maintenance on fuel economy', in D. R. Blackmore and A. Thomas (Eds), *Fuel Economy of the Gasoline Engine*, Macmillan, London

A. Baker (1984), 'Power unit progress, friction reduction'. *Automotive Engineer*, Vol. 9, no. 6, p. 11

G. J. Barnes and R. J. Donohue (1985), 'A manufacturer's view of world emissions regulations and the need for harmonization of procedures', *SAE* 850391 (also in SP-614)

C. Barrie (1986), 'Two strokes to orbit the world', *The Engineer*, 20 October 1986, pp. 26–7

P. W. Bearman (1978), 'Turbulence and ground effects', in G. Sovran, T. Morel and W. T. Mason (Eds), *Aerodynamic Drag Mechanisms*, Plenum Press, New York

Bedford (1985), 'Bedford CF Electric', *Bedford Commercial Vehicles*, Luton, England

A. Beevers (1985), 'Plastic flywheel start buses', *The Engineer*, 2 May 1985, p. 36

W. Berg (1985), 'Evolution of motor vehicle emission control legislation in Europe – leading to the catalyst car?', *SAE* 850384 (also in SP-614)

G. L. Bird (1985), 'The Ford 2.5 litre direct injection naturally aspirated diesel engine', *Proc. I. Mech. E.*, Vol. 199, No. D2, pp. 113–22

D. R. Blackmore and A. Thomas (1977), *Fuel Economy of the Gasoline Engine*, Macmillan, London

R. Buchheim, J. Maretzke and R. Piatek (1985), 'The control of aerodynamic parameters influencing vehicle dynamics', *SAE* 850279

R. Burt (1977), 'The measurement of fuel economy', in D. R. Blackmore and A. Thomas (Eds), *Fuel Economy of the Gasoline Engine*, Macmillan London

C. Campbell (1978), '*The Sports Car*', 4th edn., Chapman and Hall, London

D. Cole (1984), 'Automotive fuel economy', in J. C. Hilliard and G. S. Springer (Eds), *Fuel Economy*, Plenum Press, New York

R. M. Cole and A. C. Alkidas (1985) 'Evaluation of an air-gap-insulated piston in a divided-chamber diesel engine', *SAE* 850359 (also in *SAE* SP-610)

D. Collins and J. Stokes (1983), 'Gasoline Combustion chambers – compact or open?' *SAE* 830866

H. Daneshyar, J. M. C. Mendes-Lopes, G. S. S. Ludford and P. S. Tromans (1983), 'The influence of straining on a premixed flame and its relevance to combustion in SI engines', Paper C50/83, *Int. Conf. on Combustion in Engineering*, Vol. I, pp. 191–9, I. Mech. E. Conference Publication, MEP, London

G. O. Davies (1983), 'The preparation and combustion characteristics of coal derived transport fuels', Paper C85/83, Int. Conf. on Combustion in Engineering, Vol. II, I. Mech. E. conference publications, MEP, London

Department of Energy (1981–5), *Digest of United Kingdom Energy Statistics*, HMSO, London

L. W. DeRead (1977), 'The influence of road surface texture on tire rolling resistance', Conf. Proc. P-74, *Tire Rolling Losses and Fuel Economy*, SAE

J. A. Dominy and R. G. Dominy (1984), 'Aerodynamic influences on the performance of the Grand Prix racing car', *Proc. I. Mech. E.*, Vol. 198D, No. 12, pp. 1–7

M. A. Dorgham (ed.) (1982), *Ford Energy Report*, Technological Advances in Vehicle Design, SP1, Interscience Enterprises, Jersey, Channel Islands, UK

J. Dunn (1985), 'Out goes the distributor in Saab ignition', *The Engineer*, 24 January 1985, p. 32.

B. H. Eccleston and R. W. Hurn, 'Ambient temperature and trip length – influence on automotive fuel economy and emissions', in A. P. S. Hyde, R. F. Irwin and R. C. Stahman (Eds), *Automotive Fuel Economy Part 2*, SAE PT-18

B. D. Edwards (1984), 'Electric vehicle production and associated component developments in the United Kingdom', *SAE* 840477

J. R. Ellis (1969), *Vehicle Dynamics*, London Business Books

B. E. Enga, M. F. Buchman and I. E. Lichtenstein (1982), 'Catalytic control of diesel particulate', *SAE* 820184 (also in SAE P-107)

E. M. Evans and P. J. Zemroch (1984), 'Measurement of the aerodynamic and rolling resistances of road tanker vehicles from coast-down tests', *Proc. I. Mech. E.*, Vol. 198D, No. 11, pp. 211–18

C. R. Ferguson (1986), '*Internal Combustion Engines*', Wiley, New York

F. Forlani and E. Ferrati (1987), 'Microelectronics in electronic ignition status and evolution', *16th ISATA* pp. 17–36, Proceedings Keynote Speeches

D. J. Fothergill, R. Southall and E. Osmond (1984), 'Computer aided concept design of a sports car chassis system', Paper C182/84, *Vehicle Structures*, I. Mech. E. conference publications, 1984–7, MEP, London

R. J. Francis and L. N. Woollacott (1981), 'Prospects for improved fuel economy and fuel flexibility in road vehicles', *Energy Paper No. 45*, Department of Energy, HMSO, London

K. Garrett (1986), 'Vehicle perspective', *Automotive Engineer*, Vol. 11, No. 1, pp. 6–12

P. E. Glikin (1985), 'Fuel injection in diesel engines', *Proc. I. Mech. E.*, Vol. 199, No. 78, pp. 1–14

L. de Goncourt and K. H. Sayers (1985), 'A composite automobile front suspension', *30th National SAMPE Symposium*, pp. 708–23

E. M. Goodger (1975), *Hydrocarbon Fuels*, Macmillan, London

M. C. Goodwin and M. L. Haviland (1978), 'Fuel economy improvements in EPA and road tests with engine oil and rear axle lubricant viscosity reduction', 780596 (also in *SAE* PT-18)

C. J. Greenwood (1984), 'The design, construction and operation of a commercial vehicle continuously variable transmission', Paper C11/84, *Driveline 84*, I. Mech. E. conference publications 1984–1, MEP, London

W. W. Gregg (1983), 'GMC Aero Astro body panels', *SAE* 831003 (also in SP-545)

H. W. Hahn (1986), 'Improving the overall efficiency of trucks and buses', *Proc. I. Mech. E.*, Vol. 200, No. D1, pp. 1–13

D. Hahne (1984), 'A continuously variable automatic transmission for small front wheel drive cars', Paper C2/84, *Driveline '84*, I. Mech. E. conference publications 1984–1, MEP, London

M. S. Hancock, D. J. Buckingham and M. R. Belmont (1986), 'The influence of arc parameters on combustion in a spark-ignition engine', *SAE* 860321

E. K. Hanson (1979), 'An overall design approach to improving passenger car fuel economy', *SAE* 780132 (also in SAE PT-18)

S. Hara, Y. Nakajima and S. Nagumo (1985), 'Effects of intake valve closing timing on SI engine combustion', *SAE* 850074

N. Hay, P. M. Watt, M. J. Ormerod, G. P. Burnett, P. W. Beesley and B. A. French (1986), 'Design study for a low heat loss version of the Dover engine', *Proc. I. Mech. E.*, Vol. 200, No. DI, pp. 53–60

R. A. Haywood (1972), *Thermodynamic Tables in SI (Metric) Units*, 2nd edn. CUP, Cambridge

K. L. Hoag, M. C. Brands and W. Bryzik (1985), 'Cummins/TACOM adiabatic engine program', *SAE* 850356 (also in SAE-SP-610)

C. T. Hoffman and D. J. Beurmann (1979), 'Measurement and reduction of on-road brake drag', *SAE* 790723 (also in SAE SP-452)

W. H. Hucho (1978), 'The aerodynamic drag of cars', in G. Sovran, T. Morel and W. T. Mason (Eds), *Aerodynamic Drag Mechanisms*, Plenum Press, New York

W. H. Hucho, L. J. Janssen and H. J. Emmelmann (1976), 'The optimisation of body details – a method for reducing the aerodynamic drag of road vehicles', *SAE* 760185 (also in SAE PT-18)

J. M. Ironside and P. W. R. Stubbs (1981), 'Microcomputer control of an automatic Perbury transmission', Paper C200/81, *3rd Int. Automotive Electronics Conf.*, I. Mech. E. conference publications, MEP, London

Isuzu (1986), 'Variable intake swirl and variable geometry turbo', *Automotive Engineer*, Vol. 11, No. 3, p. 38

M. Jacobson (1984), 'Body construction techniques, steel forming', *Automotive Engineer*, Vol. 9, No. 4, pp. 42–50

R. P. Jarvis (1984), 'Fuel economy with small automatic transmissions', Paper C431/84, *VECON '84 Fuel Efficient Power Trains and Vehicles*, I. Mech. E. conference publications 1984–14, MEP, London

P. Jefferson (1985), 'Lean burn – the rational alternative to catalysts', *Motor*, 4 May 1985, pp. 40–3

R. A. Johnson (1984), 'The AGT 100 automotive gas turbine', *Mechanical Engineering*, Vol. 106, No. 5, pp. 36–43

W. K. Klamp (1977), 'Power consumption of tires related to how they are used', Conf. Proc. p-74, *Tire Rolling Losses and Fuel Economy*, SAE

G. L. Knighton (1984), 'Total vehicle economy – the challenge (United States of America)', Paper C452/84, *VECON '84 Fuel Efficient Power Trains and Vehicles*, I. Mech. E. conference publications, 1984–14, MEP, London

E. A. Koivunen and P. A. LeBar (1979), 'A new automatic transmission for improved fuel economy – General Motors THM125', *SAE* 790725 (also in SP-452)

J. T. Kummer (1984), 'Fuel economy and emissions', in J. C. Hilliard and G. S. Springer (Eds), *Fuel Economy*, Plenum Press, New York

N. Ladommatos and C. R. Stone (1986), 'Developments for direct injection diesel engines', *I. Mech. E. Seminar on Practical Limits of Efficiency of Engines*, pp. 41–53

W. Lees (1985), *Adhesives in Engineering Design*, The Design Council, London

F. W. Lohr (1984), 'Total vehicle economy – the challenge (Europe)', Paper C451/84, *VECON '84 Fuel Efficient Power Trains and Vehicles*, I. Mech. E. conference publications 1984–14, MEP, London

K. Lorenz and K. Peterreins (1984), 'Fuel economy and performance – effects on power transmission', Paper C13/84, *Driveline '84,* I. Mech. E. conference publications 1984–1, MEP, London

J. R. Luichini (ed.) (1983), *The Rolling Resistance of Highway Truck Tires*, SAE SP-546

T. H. Ma (1986), 'Recent advances in variable valve timing', *Int. Symp. on Alternative Engines and Advanced Automatics, Vancouver, Canada, August 11–12*

R. Maly (1984), Spark ignition: its physics and effect on the internal combustion engine', in J. C. Hilliard and G. S. Springer (Eds), *Fuel Economy*, Plenum Press, New York

W. T. Mason and P. S. Beebe (1978), 'The drag related flow field characteristics of trucks and buses', in G. Sovran, T. Morel and W. T. Mason (Eds), *Aerodynamic Drag Mechanisms*, Plenum Press, New York

B. S. Massey (1983), *Mechanics of Fluids*, 5th edn, Van Nostrand Reinhold, New York

W. Merzkirch (1974), *Flow Visualisation*, Academic Press, New York

E. W. Meyer, R. Green and M. H. Cops (1984), 'Austin–Rover Montego programmed ignition system', Paper C446/84, *VECON '84 Fuel Efficient Power Trains and Vehicles*, I. Mech. E. conference publications 1984–14, MEP, London

E. Moeller (1951), 'Luftwiderstandsmessungen am VW-Lieferwagen', *Automobiltechnische Zeitschrift*, Vol. 53, No. 6, pp. 153–6

T. E. Murphy (1985), 'Power system optimisation for passenger cars', *SAE* 850030

J. D. Murrell (1979), 'Light duty automotive fuel economy... trends through 1979', in A. P. S. Hyde, R. F. Irwin and R. C. Strahman (Eds), *Automotive Fuel Economy Part 2*, SAE PT-18

K. Newton, W. Steeds and T. K. Garret (1983), *The Motor Vehicle*, 10th edn, Butterworth, London

G. Onion and L. B. Bodo (1983), 'Oxygenate fuels for diesel engines: a survey of worldwide activities', *Biomass*, Vol. 3, pp. 77–133

E. L. Padmore (1977), 'The effect of transmission lubricants on fuel economy', in D. R. Blackmore and A. Thomas (Eds), *Fuel Economy of the Gasoline Engine*, Macmillan, London

F. H. Palmer (1986), 'Vehicle performance of gasoline containing oxygenates', Paper C319/86, *Int. Conf. on Petroleum Based Fuels and Automotive Applications*, I. Mech. E. conference publications 1986–11, MEP, London, pp. 33–46.

R. C. Pankhurst and D. W. Holder (1952), '*Wind Tunnel Technique*', Pitman, London

D. A. Parker (1985), 'Ceramics technology – application to engine components', *Proc. I. Mech. E.*, Vol. 199, No. 84, pp. 1–16

J. C. Paul and J. G. LaFond (1983), 'Analysis and design of automobile forebodies using potential flow theory and a boundary layer separation criterion', *SAE* 830999 (also in SP-545)

T. C. Pearce and M. H. L. Waters (1980), 'Cold start fuel consumption of a diesel and a petrol car', *TRRL Supplementary Report 636*

L. D. Peterson and T. C. Holka (1983), 'Engineering development of the Probe IV advanced concept vehicle', *SAE* 831002 (also in SP-545)

R. Pischinger and W. Cartellieri (1972), 'Combustion system parameters and their effect upon diesel engine exhaust emissions', *SAE Trans.*, Vol. 81, Paper 720756

F. C. Porter (1979), 'Design for fuel economy – the new GM front drive cars', SAE 790721 (also in SP-452)

K. Radermacher (1982), 'The BMW Eta engine concept', *Proc. I. Mech. E.*, Vol. 196, No. 9, pp. 95–103

M. L. P. Rhodes and D. A. Parker (1984), 'AEconoguide – the low friction piston', *SAE* 840181

R. M. Santer and M. E. Gleason (1983), 'The aerodynamic development of the Probe IV advanced concept vehicle', *SAE* 831000 (also in SP-545)

H. Schlichting (1960), 'Boundary layer theory', McGraw-Hill, New York

D. Scott (1986a), 'Beta battery spurred toward volume production', *Automotive Engineering*, Vol. 94, No. 1, pp. 46–7

D. Scott (1986b), 'Low-drag van built in 36 variants', *Automotive Engineering*, Vol. 94, No. 4, pp. 80–4

A. H. Seilly (1981), 'Colenoid actuators – further developments in extremely fast acting solenoids', *SAE* 810462

D. E. Seizinger, W. F. Marshall and A. L. Brooks (1985), 'Fuel influences on diesel particulates', *SAE* 850546

Shell (1986), Additives in gasoline, *Shell Science and Technology*, Issue No. 5, April

G. Sovran (1978), in G. Sovran, T. Morel and W. T. Mason (Eds), *Aerodynamic Drag Mechanisms*, Plenum Press, New York

R. Stone (1985), '*Introduction to Internal Combustion Engines*, Macmillan, London

C. R. Stone and D. I. Green-Armytage (1987), 'Comparison of methods for the calculation of mass fraction burnt from engine pressure time diagrams', *Proc. I. Mech. E.*, Vol. 201, No. DI

C. R. Stone and E. K. M. Kwan (1985), 'Variable valve timing for IC engines', *Automotive Engineer*, Vol. 10, No. 4, pp. 54–8

K. Takeuchi, K. Kuboba, M. Konagai, M. Watanabe and R. Kihara (1985), 'The new Isuzu 2.5 liter 4-cylinder direct injection diesel engine', *SAE* 850261 (also in SP-615)

C. F. Taylor (1966), *The Internal Combustion Engine in Theory and Practice*, Vol. 1, MIT Press, Massachusetts

G. D. Thompson and M. Torres (1977), 'Variations in tire rolling resistance – 'a real world' information need', Conf. Proc. P-74, *Tire Rolling Losses and Fuel Economy*, SAE

R. H. Thring (1981), 'Engine transmission matching', *SAE* 810446

G. B. Toft (1984), 'A cold-start track test procedure for evaluating fuel-efficient oils', Paper C424/84, *VECON '84 Fuel Efficient Power Trains and Vehicles*, I. Mech. E. conference publications 1984–14, MEP, London

G. Torazza (1972), 'A variable lift and event control device for piston engine valve operation', Paper 2/10, *14th FISITA Congress*, pp. 59–67

W. R. Wade, P. H. Havstad, E. J. Ounsted, F. H. Trinkler and I. J. Garwin (1984), 'Fuel economy opportunities with an uncooled DI diesel engine', *SAE* 841286 (also in SP-610)

F. J. Wallace, M. Tarabad and D. Howarth (1983), 'The differential compound engine – a new integrated engine transmission system for heavy vehicles', *Proc. I. Mech. E.*, Vol. 197A, pp. 1–11

P. Walzer, H. Heirrich and M. Langer (1985), 'Ceramic components in passenger-car diesel engines', *SAE* 850567

D. Ward (1985), 'Steady at last!', *Motor*, 12 January 1985, pp. 14–16

N. Watson (1983), 'Resonant intake and variable geometry turbocharging systems for a V8 diesel engine', *Proc. I. Mech. E.*, Vol. 197A, pp. 27–34

N. Watson and M. S. Janota (1982), *Turbochanging the Internal Combustion Engine*, Macmillan, London

G. G. Webb (1984), 'Torsional stiffness of passenger cars', Paper C172/84, *Vehicle Structures*, I. Mech. E. conference publications 1984–7, MEP, London

F. J. Weinberg (1983), 'Plasma jets in combustion', Paper C85/3, *Int. Conf. on Combustion in Engineering*, Vol. I, MEP, London

J. Yamaguchi (1986a), 'Honda chooses 90 degree V-6 to lower hood height', *Automotive Engineering*, Vol. 94, No. 2, pp. 125–31

J. Yamaguchi (1986b), 'Optical sensor feeds back diesel ignition timing', *Automotive Engineering*, Vol. 94, No. 4, pp. 84–5

J. Yamaguchi (1986c), 'Quad cam V-6 adopts variable valve timing and induction', *Automotive Engineering*, Vol. 94, No. 6, pp. 102–5

Index